Digital Technology Enabled Circular Economy

This book presents cutting-edge findings that draw on the use of AI, the Industrial Internet of Things, Blockchain, and Co-Analytics for the development of Circular Economy (CE) models to make organizational activities more sustainable. A further goal is the development of Digital Technology (DT)–enabled support tools that can be used to further analyze the impact of DT and CE–enabled operational practices used to achieve resource and environmental sustainability.

Digital Technology Enabled Circular Economy: Models for Environmental and Resource Sustainability discusses the integration of digital technology-enabled circular economy models into the manufacturing industries and its advantages for sustainability. It emphasizes the fundamentals and applications and their enactment, as well as integration for the overall organizational development. The book explains the role of digital technologies in food supply chains and multi-life cycle materials for sustainable development and highlights the development of the workforce to facilitate the implementation of smart and advanced technologies. This book presents the development of DT–enabled support tools used to analyze the impact of DT and CE–enabled operational practices on an organization and efforts to achieve resource and environmental sustainability. Case studies that demonstrate how smart digital technology can help firms improve their long-term performance by embracing circular operation methods are also included.

Students, academicians, and researchers, as well as managers and stakeholders who are interested in smart, sustainable production, and consumption, together with managers and stakeholders who are interested in ways of implementing them in their organizations, will find this book of interest. It will demonstrate, via the use of real-world case studies, how smart digital technology can help firms improve their long-term performance by embracing circular operating methods.

Advances in Intelligent Decision-Making, Systems Engineering, and Project Management

This new book series will report the latest research and developments in the field of information technology, engineering and manufacturing, construction, consulting, healthcare, military applications, production, networks, traffic management, crisis response, human interfaces, and other related and applied fields. It will cover all project types, such as organizational development, strategy, product development, engineer-to-order manufacturing, infrastructure and systems delivery, and industries and industry-sectors where projects take place, such as professional services, and the public sector including international development and cooperation, and so on. This new series will publish research on all fields of information technology, engineering, and manufacturing including the growth and testing of new computational methods, the management and analysis of different types of data, and the implementation of novel engineering applications in all areas of information technology and engineering. It will also publish on inventive treatment methodologies, diagnosis tools and techniques, and the best practices for managers, practitioners, and consultants in a wide range of organizations and fields including police, defense, procurement, communications, transport, management, electrical, electronic, aerospace, and requirements.

Hybrid Intelligence for Smart Grid Systems
Edited by Seelam VSV Prabhu Deva Kumar, Shyam Akashe, Hee-Je Kim, and Chinmay Chakrabarty

Machine Learning-Based Fault Diagnosis for Industrial Engineering Systems
Rui Yang and Maiying Zhong

Smart Technologies for Improved Performance of Manufacturing Systems and Services
Edited by Bikash Chandra Behera, Bikash Ranjan Moharana, Kamalakanta Muduli, and Sardar M. N. Islam

Wireless Communication Technologies: Roles, Responsibilities and Impact of IoT, 6G, and Blockchain Practices
Edited by Vandana Sharma, Balamurugan Balusamy, Gianluigi Ferrari, and Prerna Ajmani

Digital Technology Enabled Circular Economy: Models for Environmental and Resource Sustainability
Edited by Bikash Ranjan Moharana, Bikash Chandra Behera, and Kamalakanta Muduli

For more information about this series, please visit: https://www.routledge.com/Advances-in-Intelligent-Decision-Making-Systems-Engineering-and-Project-Management/book-series/CRCAIDMSEPM

Digital Technology Enabled Circular Economy

Models for Environmental and Resource Sustainability

Edited by
Bikash Ranjan Moharana,
Bikash Chandra Behera, and
Kamalakanta Muduli

CRC Press is an imprint of the
Taylor & Francis Group, an **informa** business

Designed cover image: Shutterstock - ImageFlow

First edition published 2025
by CRC Press
2385 NW Executive Center Drive, Suite 320, Boca Raton FL 33431

and by CRC Press
4 Park Square, Milton Park, Abingdon, Oxon, OX14 4RN

CRC Press is an imprint of Taylor & Francis Group, LLC

Library of Congress Cataloging-in-Publication Data
Names: Moharana, Bikash, editor. | Behera, Bikash Chandra, editor. |
Muduli, Kamalakanta, 1977- editor.
Title: Digital technology enabled circular economy : models for
environmental and resource sustainability / edited by Bikash Moharana,
Bikash Behera and Kamalakanta Muduli.
Description: First edition. | Boca Raton, FL : CRC Press, 2024. | Includes
bibliographical references and index.
Identifiers: LCCN 2024003713 (print) | LCCN 2024003714 (ebook) |
ISBN 9781032392493 (hbk) | ISBN 9781032394732 (pbk) | ISBN 9781003349877 (ebk)
Subjects: LCSH: Business--Data processing. | Sustainable development--Data
processing. | Blockchains (Databases)
Classification: LCC HF5548.2 .D523 2024 (print) | LCC HF5548.2 (ebook) |
DDC 338/.064--dc23/eng/20240412
LC record available at https://lccn.loc.gov/2024003713
LC ebook record available at https://lccn.loc.gov/2024003714

ISBN: 978-1-032-39249-3 (hbk)
ISBN: 978-1-032-39473-2 (pbk)
ISBN: 978-1-003-34987-7 (ebk)

DOI: 10.1201/9781003349877

Access the Support Material: Routledge.com/ 9781032392493

Typeset in Times
by SPi Technologies India Pvt Ltd (Straive)

Contents

Preface

This book focuses on various digital technologies and how they might enhance circular economy policies and practices by enhancing their good repercussions and contributing to the attainment of the environmentally Sustainable Development Goals (SDGs). As a result of the necessity to adapt to new technology and business models, these seismic shifts may result in the creation of new jobs as well as the redistribution of existing occupations throughout the world. The incorporation of digital technologies has on the other hand resulted in a considerable rise in decision complexity, particularly in the areas of ecofriendly product and process design, refurbishment, remanufacturing, and logistics procedures, which are considered as some of the prominent activities adopted by many organizations as part of their environmental and resource sustainability programs. Waste, resource consumption, and other environmental consequences are reduced to a basic minimum all through the life cycle of products and services, and at the same time acknowledging socioeconomic progress is achieved through the use of this innovative concept.

The book *Digital Technology Enabled Circular Economy Models for Environmental and Resource Sustainability* comprises of 15 chapters, which discusses the integration of digital technologies for socioeconomic progress in several sectors including energy sectors at first and followed by occupational health hazards, circular economy business models, sustainable manufacturing and materials system, supply chain management, logistics and marketing strategies, and finally a sustainable cooling technique for machining industries. This book offers a wide range of discussions that are complemented by research insights into the areas of recent development on different digital technologies and their circular economy policies for overall sustainable development, as well as real-time case studies that showcase the implementations of smart digital technology.

The first chapter focuses on sustainable energy generation and includes the roles and applications of digital technologies in solar systems and the multiple business opportunities that they offer for better understanding and to help develop a digital mindset among solar power stakeholder groups. Also, the exploration of some limitations of digitalization in the "solar energy" sector and possible solutions are discussed. All of these will be elaborated on through multiple case studies for better understanding.

The second chapter explains the latest advancements in solar still technology, which could help the society to generate clean and drinkable water from brackish and saline water sources. This technology would have significant advantage particularly in arid and semiarid regions, where the limited availability of drinking water is more pronounced compared to other places on the globe.

The third chapter focuses on the ergonomic assessment through a sustainable statistical approach. The analysis may be helpful to assess the impact of occupational health hazards in order to optimize human well-being and overall system performance. It may provide awareness for the laboratory demonstrators and machine operators in educational institutions.

The fourth chapter presents the impact of combining circular economy models with digital technologies on employment generation and redistribution. AI and IoT enhance CE models by optimizing resource utilization, enabling proactive maintenance, and facilitating streamlined supply chain management. This optimization often results in novel employment opportunities in fields such as data analysis, IoT device administration, and sustainability consultancy.

The fifth chapter gives a brief discussion on an EOQ model for deteriorating items. An optimal ordering policy for deteriorating items is proposed by incorporating the dynamic factors such as variable deterioration and exponentially declining demand and by ending with shortages over a fixed time horizon, which minimizes the cost function of the inventory system.

The sixth chapter focuses towards the sustainability indicators for smart and connected digital manufacturing systems. It may be helpful for the manufacturers to reduce their carbon footprint, conserve resources, and ensure the well-being of their workers and local communities by evaluating the environmental and social effects of digital manufacturing systems.

The seventh chapter highlights the concept and importance of environmental sustainability in the coming time via a case study on robots. This presents a challenge and an opportunity for young researchers worldwide to contribute towards the development of sustainable robots, which is a critical factor in achieving a more sustainable future.

The eighth chapter provides the recent advancements in sustainable fabric materials and digitally enhanced textiles, including 3D printing and digital twinning employing DCNN. It also offers an invaluable insight and paves the way for future developments in enhancing comfort and safety in smart textiles for firefighter clothing.

The ninth chapter highlights the importance of IIoT in enabling real-time monitoring through continual sensor-based data gathering, assuring perfect conditions and food safety. Smart warehousing powered by IIoT transforms perishable food supply chains, promoting sustainability via decreased food waste, increased energy efficiency, and responsible resource utilization, transforming the industry's future.

The tenth chapter explains the impact of green marketing strategies on consumer attitudes towards environmental sustainability. A business model requires significant advancements in eco-labelling as well as environmental concerns and attitudes in order to develop green brands that contribute to the long-term viability of the environment.

The eleventh chapter presents the influence of digital technology and sustainable smart product-service systems on greenhouse gas emissions. The framework examines digital technologies, such as IoT sensors, data analytics, and AI-driven optimisation, emphasizing their integration into PSS and their impact on emissions reduction, resource conservation, and energy efficiency.

The twelfth chapter provides the connections between green logistics practise, digital technologies and adaptation of circular economy, and its importance to promoting sustainable development. The opportunities, challenges, and suggestions, including the implications of sustainability and development practices in logistics on industry people, academics, and researches are highlighted as a way forward for us.

The thirteenth chapter focuses on AI-driven technologies for the creation of advertising content, which can produce graphics, movies, and ad copy that are optimized for specific audiences by developing innovative practices oriented towards sustainability and the creation of new markets.

The fourteenth chapter explains about an alternative sustainable cooling technique for machining industries, that is, Minimum Quantity Lubrication (MQL). It discusses the adverse effects associated with flood cooling and explores the advantages of MQL as a sustainable cooling and lubrication method.

The concepts discussed in this book will be very much supportive and encourage the readers, including enthusiastic researchers, industrialists, scientists, academicians, and undergraduate and postgraduate students, to understand and explore various smart technologies enabled circular economy in the milieu of sustainability. We express our gratitude to all of the authors who have contributed materials for this book as well as to the readers who will be using it.

Bikash Ranjan Moharana
PNG University of Technology, Lae
Bikash Chandra Behera
C V Raman Global University, Odisha
Kamalakanta Muduli
PNG University of Technology, Lae

Editors

Dr. Bikash Ranjan Moharana is presently working as senior lecturer in the School of Mechanical Engineering, Papua New Guinea University of Technology, Lae, Morobe Province, Papua New Guinea. He received his Ph.D. in Mechanical Engineering from the National Institute of Technology, Rourkela, India in 2018. He hobtained his master's degree from Veer Surendra Sai University of Technology, Burla, Odisha. He has published more than 25 papers in various peer reviewed Scopus and SCI indexed journals, conference proceedings, and book chapters. He has been guest editing two books approved for publication by Taylor & Francis Group, CRC Press. His main research interests focus on various fusion welding processes, non-traditional machining, process optimization, and mechanical and metallurgical analysis. He is a fellow member in Institution of Engineers India, IIW and IWS.

Dr. Bikash Chandra Behera is an assistant professor at the Department of Mechanical Engineering, C. V. Raman Global University, Bhubaneswar, India. His research interests include sustainable machining, machining tribology and micro machining. He is currently working in the area of artificial intelligence and machine learning applications in manufacturing processes. Dr. Behera received his master's degree and doctoral degree from the NIT, Rourkela, India and the Indian Institute of Technology Delhi, India, respectively. Over the course of his career, he has authored more than 30 articles in peer-reviewed reputable journals, has presented articles at globally reputed conferences, and has published book chapters.

Dr. Kamalakanta Muduli is presently working as an associate professor in the School of Mechanical Engineering, Papua New Guinea University of Technology, Lae, Morobe Province, Papua New Guinea. He has obtained a Ph.D. from the School of Mechanical Sciences, IIT Bhubaneswar, Orissa, India. He has obtained a master's degree in industrial engineering. Dr. Muduli has over 15 years of academic experience in universities in India and Papua New Guinea. Dr. Muduli is a recipient of the ERASMUS+ KA107 award provided by the European Union. He has published 54 papers in peer reviewed international journals, most of which are indexed in Clarivate analytics, Scopus and listed in ABDC and more than 25 papers in national and international conferences. He has been also guest editing few special issues in journals and books approved for publication by Taylor and Francis Group, MDPI, CRC Press, Wiley scrivener, and Apple Academic Press. Dr. Muduli also has guided three Ph.D. students. His current research interest includes materials science, manufacturing, sustainable supply chain management, and Industry 4.0 applications in operations and supply chain management. Dr. Muduli is a fellow of the Institution of Engineers India. He is also a senior member of the Indian Institution of Industrial Engineering and member of ASME.

Contributors

Jnanaranjan Acharya
National Institute of Technology
Silchar, India

Bikash Chandra Behera
C. V. Raman Global University
Bhubaneswar, India

Dipankar Bhanja
National Institute of Technology
Silchar, India

Dillip Kumar Biswal
Krupajal Engineering College
Bhubaneswar, India

Jitendra Narayan Biswal
Einstein Academy of Technology and Management
Bhubaneswar, India

B. S. Chawla
Govt. Engineering College
Bilaspur, India

Shailesh Dewangan
Chouksey Engineering College
Bilaspur, India

Patrick Dichabeng
University of Botswana
Gaborone, Botswana

Granville Embia
Papua New Guinea University of Technology
Lae, Papua New Guinea

Manoj Kumar Gopaliya
The NorthCap University
Gurugram, India

Shiv Manjaree Gopaliya
VIT Bhopal University
Kothri Kalan, India

Pushpdant Jain
VIT Bhopal University
Kothri Kalan, India

Modibo Kante
JAIN (Deemed to be University)
Bengaluru, India

Virendra Kumar
National Institute of Technology
Silchar, India

Benson Mirou
Papua New Guinea University of Technology
Lae, Papua New Guinea

Rahul Dev Misra
National Institute of Technology
Silchar, India

Richie Moalosi
University of Botswana
Gaborone, Botswana

Sephali Mohanty
C. V. Raman Global University
Bhubaneswar, India

Bikash Ranjan Moharana
Papua New Guinea University of Technology
Lae, Papua New Guinea

Tapas Kumar Moharana
National Council of Science Museums
Kolkata, India

Kamalakanta Muduli
Papua New Guinea University of Technology
Lae, Papua New Guinea

Noorhafiza Muhammad
Universiti Malaysia Perlis
Perlis, Arau, Malaysia

Arpita Nayak
KIIT University
Bhubaneswar, India

David Chua Sing Ngie
Universiti Malaysia Sarawak
Kota Samarahan, Sarawak, Malaysia

Atmika Patnaik
King's College
London, United Kingdom

B.C.M. Patnaik
KIIT University
Bhubaneswar, India

Ashutosh Pattanaik
JAIN (Deemed to be University)
Bengaluru, Karnataka, India

Adimuthu Ramasamy
Papua New Guinea University of Technology
Lae, Papua New Guinea

Yaone Rapitsenyane
University of Botswana
Gaborone, Botswana

Manidatta Ray
Birla Global University
Bhubaneswar, India

Victor Ruele
University of Botswana
Gaborone, Botswana

Suchetana Sadhukhan
VIT Bhopal University
Kothri Kalan, India

Ipseeta Satpathy
KIIT University
Bhubaneswar, India

Oanthata Sealetsa
University of Botswana
Gaborone, Botswana

FonoTamo Romeo Sephyrin
University of Tennessee
Knoxville, Tennessee, U.S.A.

Keiphe Setlhatlhanyo
University of Botswana
VGaborone, Botswana

Arun Kumar Singh
Papua New Guinea University of Technology
Lae, Papua New Guinea

Trailokyanath Singh
C. V. Raman Global University
Bhubaneswar, Odisha, India

Chitrakant Tiger
Chouksey Engineering College
Bilaspur, Chhattisgarh, India

Prabhuram Tripathy
Sri Sri University
Odisha, India

Sushanta Tripathy
KIIT University
Bhubaneswar, India

1 An Insight of Digital Technology's Role in the Solar Energy Sector and Its Implications for Sustainable Energy Generation

Modibo Kante and Ashutosh Pattanaik
JAIN (Deemed to be University), Bengaluru, India

Bikash Ranjan Moharana
Papua New Guinea University of Technology, Lae, Papua New Guinea

1.1 INTRODUCTION

Sunlight, tides, rain, wind, and geothermal energy are the primary sources of sustainable energy; they are naturally replenished and do not deplete over time. In contemporary times, fossil fuels have emerged as the predominant primary energy source, although their availability is gradually diminishing as they are being depleted from the Earth's storage. In pursuit of ensuring sustained satisfaction of their fundamental energy requirements, individuals globally have been actively exploring alternative sources to enhance energy resilience. Environmental pollution increases with the fast growing of the population and the growing consumption of fossil fuels. Consequently, there exists an acute need for the implementation of sustainable and environmentally friendly practices, which have gained widespread acceptance among nations around the globe. As the renewable energy sources are clean and there is no pollution consumption, they are attractive to the entire globe, and hence a large capital investment is being made in harvesting these resources. Decarbonisation is becoming the primary goal of the entire globe, as the use of non-renewable energy sources degrades the planet continuously. Renewable energy is expected to be the source of energy that has the greatest percentage increase in global demand over the next few decades. In spite of the growth in renewable energy sources, it is anticipated that coal, oil, and natural gas will continue to hold their positions as the primary sources of energy by

DOI: 10.1201/9781003349877-1

the year 2040, and 77% of the world's energy demand will still be met by fossil fuels at that time. It is anticipated that natural gas will be the form of fossil fuel that has the greatest rate of growth.

In 2017, petroleum and liquid fuels contributed 33% of the world's energy consumption; however, this percentage is anticipated to drop to 31% by 2040, while overall energy consumption is forecast to rise by 28% by the same year (Taşaltin, 2019). According to a report by the International Solar Energy Association and Wood Mackenzie Power and Renewables, the number of solar power installations in the United States had exceeded 2.2 million by the year 2020. Simultaneously, it is projected that the number of solar power plants in the United States will reach 4 million by the year 2023. China now holds the position of global frontrunner in the realm of electricity generation derived from renewable energy sources. As of the commencement of 2020, China's power generating capacity reached over 800 GW, surpassing that of the United States, which occupies the second position worldwide, by a factor of two. Hence, the effective control of temporal consumption emerges as a crucial instrument for facilitating the energy sector's transition towards renewable electricity programmes (Valeeva et al., 2022). In the pursuit of addressing climate change, it is imperative to direct attention towards not just augmenting alternate energy sources, but also towards diminishing energy use and formulating strategies for enhancing energy efficiency (Taşaltin, 2019).

Digital technologies are enabling the development of digital innovations through the utilisation of various tools, technologies, and processes, which includes big data analysis, blockchain technology, artificial intelligence & machine learning (AI & ML), Internet of Things, robotics, sensors, and so on. These innovations are motivating stakeholders in the energy sector to explore novel opportunities that enable the generation, delivery, and consumption of energy in a sustainable manner. These novel technologies are also facilitating the emergence of fresh revenue-generating prospects through inventive company structures.

Digitisation of energy firms enables them to effectively decrease operational expenses and enhance overall efficiency. The implementation of digital advancements has the potential to extend the operational lifespan of an energy plant by up to 30%. The integration of blockchain technologies together with the use of intelligent networks has facilitated the engagement of consumers within the electrical market (Borowski, 2021). The utilisation of artificial intelligence (AI) methodologies in the context of power and renewable energy systems has been a prominent subject of academic research for over three decades, yielding noteworthy accomplishments. Artificial neural networks (ANN), fuzzy logic, and expert/knowledge-based systems have emerged as the most effective AI approaches. AI approaches are of significant importance in the modelling, analysis, prediction, performance evaluation, control, and safety assurance of renewable energy systems. AI is also used in the formulation of the forecasting and estimation model of the solar meteorological data (Belu, 2013; Kalogirou and Sencan, 2010). The most effective approach to addressing this issue is to implement remote monitoring of systems through the use of the developing concept known as the Internet of Things (IoT). In the realm of the IoT, various information and communication technology devices are seamlessly integrated into the internet infrastructure through the utilisation of microcontrollers, various digital communication tools, information and network protocols, and so on. The technology has the potential to facilitate the comprehensive collection of information about things, hence offering a wide range of new opportunities for growth (Wu et al., 2022). The application of sector 4.0 to photovoltaic systems enhances their stability and dependability, hence enabling more efficient management of energy production. This integration has the potential to enhance the competitiveness of the solar sector. Through the utilisation of IoT technology, energy managers are able to remotely monitor many photovoltaic systems in real time, regardless of their location, by connecting to the internet via personal computers or smartphones. This enables them to promptly detect and analyse the underlying causes of defects and issues inside these systems. The utilisation of real-time monitoring technology in solar systems enables seamless maintenance procedures, hence obviating the necessity for costly and arduous unscheduled on-site testing. The utilisation of sensor technology enables the solar panels to accurately detect the sun's movement and adjust their orientation accordingly, optimising their effectiveness by maximising exposure to solar radiation. The utilisation of blockchain technology is believed to enhance the efficiency of energy exchange within electrical transmission systems. This method allows for the inclusion of consumer units as producers inside the system. By use of this system, it is possible for each unit to monitor the energy generated in the other unit, resulting in a transparent system. In the context of solar energy adoption, a residential property equipped with a solar panel system can participate in the market by engaging in peer-to-peer transactions, wherein surplus energy generated by the house is sold to individual consumers rather than centralised entities. Therefore, the establishment of a local solar market might lead to the assurance of money flow. The excess energy generated by photovoltaic panels installed on the rooftops of structures has the potential to be commercially traded with adjacent buildings. Despite the interconnectedness of these structures through the urban network, the use of blockchain technologies for commercial transactions is feasible. This technological advancement enables the monitoring of real-time energy production levels, associated costs, geographical locations of energy generation, and the corresponding energy demand locations. Big-data analytics is a prominent field within AI that has made significant progress in the digital realm. In the context of renewable energy, sustainability aspects such as decision-making, planning, status monitoring, maintenance, robotics, audits, and supply chain optimisation can be effectively applied.

The digitisation of the solar power sector encompasses technologies that digitally monitor, manage, and optimise energy generation (Kangas et al., 2021). These technologies also open up new avenues for more efficient data collection, management, and analysis (Pakulska and Poniatowska-Jaksch, 2022). They create several business opportunities in the global market.

With this background, the present study was conducted to answer the following research questions.

RQ1: How effectively is digital technology surging in the solar energy sector, like in multiple other industries?

RQ2: The involvement of various information technology methodologies in renewable energy systems; is the digital revolution the solution or a new source of the problem?

RQ3: How is the introduction of digitalisation in the solar energy sector impacting the solar market and opening new market business?

To answer these research questions, the study was conducted with the following objectives.

- A thorough examination of the functions and uses of different digital technologies in the solar energy industry.
- A few case studies of digital technology adoption in the solar energy industry were also taken into consideration for the analysis.

The chapter is prepared as follows. Section 1.2 covers the literature review on implementation of digital technologies in the solar energy industry to improve the sustainability. Section 1.3 presents the literature gap followed by the role and applications of digital technologies in the solar energy sector in Section 1.4. Research methodology and case studies of the implementation of digital technologies are explained in Sections 1.5 and 1.7 respectively and Section 1.7 presents the result discussion. In Section 1.8 implications of this research and in Section 1.9 conclusions and future research opportunities are presented.

1.2 LITERATURE REVIEW

This section emphasises the various aspects of sustainable energy and the implementation of various digital technologies to find a solution to the energy problems through thorough review of several available researches.

1.2.1 Need of Sustainable Energy/Renewable Energy

Sustainable energy is defined as any energy source that cannot be depleted and will be viable indefinitely. It does not need to be renewed or replenished; sustainable energy meets our energy requirements without the risk of degradation or depletion. This is why renewable energy is the key to our energy issues, that is, the current energy crisis, what we are facing nowadays (Ramjeawon, 2020). Sustainable energy, for example solar/wind energy, which may be responsible for decrease of carbon footprint and contribute in a positive manner to global warming, also makes a difference in overall electric energy cost for all consumers. How renewable energy is decisively impacting environmental sustainability in many developed countries like France, Thailand, China, India, and others was discussed by many researchers (Azam et al., 2023; Abbasi et al., 2021; Nwaiwu, 2021; Sahoo, 2016; Muduli et al., 2016). As per the report from Chen et al. (2023), an approximately 40% reduction in greenhouse gases may be possible through enhancement in energy efficiency, which simultaneously attains both economic and environmental goals, leading to overall economic growth. Several researchers around the world are working on sustainable practices to reduce the carbon footprint in different ways (Biswal et al., 2017; Grimaccia et al., 2017; Mangla et al., 2020; Muduli et al., 2021; Piso et al., 2023). As energy positively impacts economic growth and degrades the environment, ultimately a source of sustainable energy is highly needed for the betterment of the environment, which directly affects human health and well-being. There are plenty of things that nations can do to develop and strengthen the environment for large-scale investments in renewable energy and sustainable technology. All nations must design their own energy strategies based on their individual requirements and available resources. For easy assessment of the resources, how digital technology contributes is discussed in the next session.

1.2.2 Digital Technology for Sustainable Energy Generation

Sustainable energy produces little or no greenhouse gas emissions. Humans can effectively decrease the rate of climate change by committing to clean energy and using sustainable energy resources. The rise of technological development, which aims to improve the efficiency of global processes in all sectors, is what propels digital transformation. To minimise the carbon footprint through sustainability practices, digital technology is the ultimate indispensable tool. There are many areas such as green energy, smart recycling systems, smart waste management, clean mobility, circular economy, sustainable food, water systems, and so on, where the practice of digital technology may enhance the environmental sustainability (Daly, 2022). Digitalisation is offering opportunities in the energy sector that did not previously exist; potential for minimising emissions while maximising efficiency through scalable solutions leads to a more sustainable future that benefits everyone. However, traditional methods are rapidly being supplemented and supported by digital solutions that enable smart linkages to minimise emissions via digitalisation and the industrial Internet of Things (IIoT) (Felicio, 2022). Development and benefits of digitalisation in the renewal energy sector, particularly in the field of renewable energy plants building, forecast of the weather and market conditions, optimisation of renewable sources and plants, and so on, are explained by Nexus-Integra (2023) and Wang et al. (2022). The renewable energy's use has become the standard technique for new power generation capacity, and nations all over the world have set ambitious goals for themselves, particularly in the solar energy industry. As per the report, by 2025, India's renewable energy capacity is expected to increase to 280 GW, or 37% of the country's total energy supply (Quixy Editorial Team, 2023). The digital transformation in solar industries also aids many positive impacts including user-friendly customer relationship management solution, accurate requirement, real-time energy monitoring, effective inventory and material management, and so forth. On another side of the coin, the digital technology also enabled solar power to be more predictable (Watson, 2017). The global scenario of the digital technology–enabled solar energy is discussed in the next session.

1.2.3 Global Scenario of Digital Technology (DT)–Enabled Solar Energy

With the advance of the technology, digital technologies are gaining more attention due to the impact they are making in various processes and operations in all the sectors. This caught the attention of the solar energy industry, which led to the implementation of DT in solar energy to improve sustainability. Hence several research works have been done assessing the impact of the implementation of DT in the solar energy sector. Some of the most significant of those research works are as follows:

Taşaltin gave a brief introduction to the applications of the most prominent digital technologies, for instance, AI, big data analysis, blockchain technology, and so forth, being used in the solar systems and also discussed some of their traditional benefits (2019). Kangas et al. focus on smart energy transition of solar and wind power using digital information and communication technologies (2021). The article also provides technology convergence analyses of renewable energy and information communication technologies inventions based on international patent data and also suggested that the smart transition may make a positive impact on renewable energy technologies.

Many researchers are investigating various aspects of AI, in particular its benefits in the solar sector; some of the difficulties of integrating AI algorithms are also discussed. Application of various AI approaches like neural network, genetic algorithm (GA), fuzzy logic approach, and so on, on solar energy systems were studied by (Belu, 2013). His study includes various aspects such as solar radiation prediction, seizing, performances of the solar photovoltaic (PV) systems, and so forth. There was a strong recommendation that the AI techniques may overcome the issues of solar energy systems in long-term measurements (Kalogirou and Sencan, 2010). A solar forecasting model was explained by Alankrita and Srivastava (2020). They highlighted that machine learning approach may be an alternative for forecasting in hybrid renewable energy systems but also emphasised the need for more analysis on battery ageing in renewable systems. The scheduling and control problems of future renewable energy systems was addressed by Hu et al. (2022). Wu et al. (2022) developed an improved fuzzy logic model for solar power generation and applied it to PV sites in Taiwan. The model shows a remarkable achievement in terms of high-level efficiency (≈7% error only) and practicability (Hu et al., 2022). During tragic issues like COVID-19, renewable energies can function effectively, lessen the virus's harmful impacts, and promote economic growth. AI plays a vital role during such periods (Sharifi et al., 2021). Impact of AIs for renewable energy in different applications including COVID-19 pandemic, carbon-neutral transition of the electric energy systems, manufacturing sectors, and others are explained in different articles (Xie et al., 2022; Malik et al., 2022; Embia et al., 2023; Chhotaray et al., 2024). Elsaraiti and Merabet described the forecasting of solar energy by deep learning method (2022). The accuracy of the prediction model is significantly impacted by the quality of the data, especially the outliers.

Like other modern technology, the IoT is used throughout the energy industry, including supply, transmission, and demand. Numerous researchers addressed the connectivity issues in the solar system using IoT as digital technologies by means of various sensor devices; as a result better connectivity and remote control of the plants were achieved (Spanias, 2017; Kumar et al., 2018; Motlagh et al., 2020; Embia et al., 2023a; Galgal et al., 2024). The implementation of big data in order to easily store and analyse the data emanated from the plants, as well as the impact of the combination of big data with other digital technologies such as machine learning and deep learning, is highlighted (Hu et al., 2015; Jiang et al., 2016; Torres et al., 2019a; Yousif, 2021; Embia et al., 2023b). Torres et al. examined the usage of several data sources (solar power and weather data from the previous days, as well as the weather forecast for the following day), in addition to the impact on various historical window sizes. The findings demonstrate that DL scales well, gives accurate answers that are competitive, and is thus an excellent approach for big-data contexts (2019b).

Blockchain is utilised primarily in the solar energy sector for trading and in the platform used for marketing the solar power (Juszczyk and Shahzad, 2022). Themistocleous et al. implemented blockchain technology in the solar energy sector and studied its impact on it. They also concluded that the technology may interrupt the energy sector in many different ways like disruption in process, product, position and paradigm, and platform innovation (2018). Wörner et al. described the possible implementation of real blockchain-based electricity market in the real world, which may lead to future sustainable energy markets (2019). The impact of digitalisation towards the management process in energy sector markets is explained by Borowski (2021). This exposes the applications of blockchain technology in the solar energy sector and the business model it offers (Baidya et al., 2021).

The impact of digitalisation in the renewable energy sector towards sustainable future energy such as sustainable economic development was studied by many researchers (Pagliaro and Meneguzzo, 2019; Sribna et al., 2021; Barinova et al., 2021; Truong, 2022; Singh et al., 2022; Zhang et al., 2023). The results provide insight into how the energy system and digitisation are developing in tandem to create a smarter, more sustainable society. The findings are also relevant to the redesign of energy education programmes in response to changes in energy technology. Positively, startups that deal with solar energy storage stand out. Rather than technology advancements, legislative adjustments should be the driving force behind these sectors' digital transition in the future. Improving the environment for innovative business models becomes essential, particularly for startups and newcomers (Pakulska and Poniatowska-Jaksch, 2022). The resources in this area include funding and grants accessible to startups, mentors who are supported, many organisations that serve as a link between startups and industry, and several regulations that support innovation. The rate of regulatory framework adaption and market participants' capacity to create viable business plans are key factors influencing the energy transformation's stability and pace. Based upon the numerous literature survey, the research gap is explicated in the next section.

1.3 RESEARCH GAP

Despite various reviews and works highlighting the importance, role, and application of DT in the solar energy sector, several issues remain persistent and unresolved. These issues include a lack of digital mindset, a lack of digital technology skills, integration in modern societies, the high amount of power consumed by digital technologies, government policies, and regulatory issues. Because of all of those issues it is important to have a clear understanding of the impact digitalisation can have on the solar energy sector and to provide some of the alternative solutions to the limitations from various research works done. In this work we will provide some solutions to the lack of digital mindset through deeper understanding of the role and applications of some prominent digital technologies in use in the solar sector and case studies which highlight the perfect implementations of digitalisation.

1.4 ROLE AND APPLICATIONS OF MOST PROMINENT DIGITAL TECHNOLOGIES IN SOLAR ENERGY SECTOR

1.4.1 Artificial Intelligent (AI)

Artificial intelligence or AI refers to the emulation of human intelligence in machines through the application of intricate algorithms and mathematical functions. AI systems possess the ability to acquire knowledge from experience, enabling them to make informed judgements and decisions based on these acquired experiences. Consequently, AI is capable of performing tasks that traditionally necessitate human intelligence. AI has been formally characterised as the discipline encompassing the scientific and engineering principles involved in the creation of intelligent machines, especially intelligent computer programmes. Artificial intelligence is a field that encompasses several disciplines and methodologies. However, the emergence of machine learning and deep learning techniques has brought about a significant transformation in nearly all domains of the technology industry. The primary goal of AI is to help the machine to think and to make its own decisions without any human intervention. Recently AI robots have started being used for maintenance; also, AI-integrated drones incorporated with UV sensors are being used in large and remote solar power plants for fault and error detection. Since its introduction in various solar systems, AI is enhancing solar power production through the integration of the various AI algorithms in solar energy systems to improve efficiency, reliability, and safety in the solar industry; helping to identify leakage; and to understand the health and energy consumption pattern of the systems, it is also being integrated in the energy storage unit. AI systems encompass a range of domains, including expert systems, ANN, data mining, GA, fuzzy logic, and diverse hybrid systems that integrate two or more approaches (Belu, 2013).

The most prominent of those algorithms used in the solar sector with their applications are as follows:

Area of Different Solar Sectors	AI Techniques
Prediction of solar radiation & control of solar buildings	ANN, Fuzzy logic
Modelling of solar steam-generator, Modelling of a solar air heater, Characterisation of Si-crystalline PV modules	ANN
Solar cell	GA, Data Mining
Photovoltaic solar energy systems	GA, Fuzzy logic
Determination of Angstrom equation coefficients, Solar water heating systems	GA
Flat plate solar air heater and its efficiency	ANN, GA
Hybrid solar–wind system PV-diesel hybrid system	GA

In order to achieve high performance, maximum power point tracking fed with an appropriate algorithm is employed to extract the maximum power from the PV array.

Non-linearity and output power and voltage variations may be tackled using smart inverters. To lessen the variability in PV output, an inverter control system based on fuzzy logic might be employed (Alankrita and Srivastava, 2020).

1.4.2 Internet of Things (IoT)

Internet of Things or IoT is a network that connects devices that share information and data about their processes, such as the way they are used and their surroundings. This is performed using sensors embedded in the devices; data obtained are shared and stored using a common platform provided by IoT in a common language to allow the communication of the devices. The IoT is a burgeoning technological advancement that leverages the power of the Internet to establish communication among tangible entities, sometimes referred to as "things".

Choosing components that are appropriate for the intended application, including sensor devices, data storage, communication protocols, and computation techniques, is a crucial first step in the development of Internet of Things systems. IoT devices, comprising sensors, actuators, IoT gateways, or other devices, serve as integral components inside IoT systems. These devices actively participate in the cycle of data gathering, transmission, and processing. The communication protocols, which constitute the third component of the IoT platform, facilitate the exchange of data between various devices and the controllers or decision-making centres. Some examples of such technologies encompass Wi-Fi, Bluetooth, ZigBee, as well as cellular technologies like LTE-4G and 5G networks. The data storage component of the IoT platform facilitates the organisation and control of data acquired from various sensors. The fifth component of IoT systems consists of stored data that is utilised for analytical purposes. Data analytics can be conducted offline following data storage, or it can be executed in real time as real-time analytics. Data analytics is conducted to inform decision-making on the operational aspects of the application. The primary role of IoT in solar energy sector is to connect the systems into a single network, making them smart, to remotely control and facilitate the monitoring, maintenance, and detection of errors and faults in the solar systems so that quick actions can be taken (Wu et al., 2022). It also enables the recording of performance data for analytics which can be used for forecasting and predicting future power generation possibilities.

Some other applications include solar energy analytics, intelligent hybrid solar PV system, remote monitoring and control of PV systems by means of smart monitoring devices incorporated with IoT sensors, forecasting and prediction of solar energy, and real-time data transferring and communication.

1.4.3 Big-data Analytics

The term big data describes a collection of data that is so large and intricate that it cannot be stored or processed by any of the conventional data application software or tools. Analysing large amounts of data to extract knowledge that can be used to influence choices is a challenging task known as big-data analytics. It is a type of advanced analytics that entails intricate applications that are made possible by analytics systems-powered statistical algorithms and predictive models. Solar energy exhibits significant variability due to its reliance on meteorological factors such as solar radiation, cloud coverage, rainfall, and temperature. The presence of this interdependence introduces a level of ambiguity regarding the quantity of solar energy that will be produced, hence complicating the process of incorporating solar power into the electrical grid and electricity markets. Therefore, precisely forecasting the amount of solar power generated is a crucial responsibility for energy managers and electricity dealers. This is essential in order to reduce uncertainty and provide a dependable electricity supply at a reasonable cost. The availability of historical photovoltaic solar power data facilitates the utilisation of advanced computer technology and machine learning methodologies for the analysis of extensive time series datasets. Solar energy with the integration of various advanced technologies generates a large amount of data. The primary role of big data is used to store those data so that they can be easily processed in a way to derive necessary information. The next step is to prepare, process, cleanse, and analyse those data using big-data tools and framework for the estimation, simulation, prediction, and optimisation of the solar energy.

The major applications of big-data analysis include

- Prediction of the energy that can be redirected into the grid
- Getting updates regarding the performance of the solar plant
- Prediction of the maintenance of the solar plant

1.4.4 Blockchain Technology

Blockchain technology is a distributed ledger that builds trust and transparency among network participants by storing digital event transactions in a chronological sequence and sharing them in a decentralised network. It enables the transfer of digital assets or coins from one individual to another. It is also described as the collection of records linked with each other, which are strongly resistant to alteration and protected using cryptography. It has the power to alter how we do business, and its disruptive impact may be amplified by combining it with other technologies like AI or IoT (Themistocleous et al., 2018).

Its foundation consists of three primary parts: distributed networks, shared ledgers, and transactions and blocks. Blockchain applications may be categorised into three main groups according to their development stage: Blockchain 1.0, Blockchain 2.0, and Blockchain 3.0. Cryptocurrencies fall into the first group, digital protocols are represented in the second, and the stage of developing the smart contract idea is the last category. Blockchain is playing a major role in the solar energy sector as it is helping to address one of the most fundamental and persistent problems in the solar energy sector, which is the storage limitations that might overload during the peak moments of energy production, and how to deal with the excess energy. As blockchain allows energy transaction in a decentralised grid between solar energy users without any centralised grid, the solar energy users connected to the same grid can both buy and sell excess energy.

Hence blockchain plays several roles and applications in solar energy which include

- Facilitating the connection of solar energy into the grid.
- Emerging the solar energy market by providing peer-to-peer platforms for selling and buying energy among the solar users, which also helps to create local energy communities.
- Tracing the energy produced and consumed and maintaining the related records.
- Facilitating monetary transactions in the solar energy sector.
- In combination with AI and IoT, making smart batteries that can make decisions related to buying and selling the solar energy produced.

Other applications include the use of digital software such as Aurora, PVsyst, Pylon, Solar Edge site designer, PV Sol Free & Premium, PV F-chart, and Solar Labs to facilitate solar plant installation and accuracy. They are also used for accurate simulations and shading analysis.

1.5 METHODOLOGY ADOPTED FOR THE RESEARCH STUDY

As this research work is meant to address some aspect of the implementation of DT in the solar sector, the methodology adopted starts with the collection of information from previous related research works and various websites. In the next step the information collected are analysed to extract the relevant limitations of digitalisation in the solar energy sector. After the analysis several problems encountered by the integration of solar power were derived; among those are the lack of digital mindset, limitation of skilled operators, definition of regulatory policies, and so on. This research work means to address the lack of digital mindset to develop better understanding of DT in the context of solar energy through the role and applications of the most significant digital technologies currently in use in the solar power sector. For that purpose the methodology of this work was based on several reviews and works done regarding digitalisation in the energy and renewable energy sectors. Some of those significant works are "Digitalization of solar energy: a perspective" (Taşaltin, 2019), which introduces the importance and applications of digitalisation in the solar energy sector, and "Digitalisation and sustainable energy transitions in Africa" (Nwaiwu, 2021), which through case studies explained the role of digitalisation in sustainable energy transition, and the business models that it offers are explained with a qualitative methodological approach for Nigeria and South Africa in order to review and analyse the energy sectors. Several reviews addressing the execution of individual digital technologies in the "solar energy" sector were also referred.

The case study uses theoretical sampling when the goal is theory building. Examples may be selected to support new theories, replicate prior examples, or fit into certain theoretical categories. Case studies provide for practical field observation, which is especially useful for discussing the implications of digital technology in the solar energy industry (Muduli et al., 2016). These provide a strong focus on in-depth qualitative investigation and are especially helpful in providing answers to the "what", "why", and "how" questions related to the issue at hand. Because the method produces detailed contextual information that might lead to a higher degree of comprehension, it is appropriate in situations where there is a deficiency of previous knowledge. There are several other benefits to the case study approach. The following are some of the major benefits (Eisenhardt, 1989):

- It makes it possible for the researchers to create grounded hypotheses that are applicable and useful.
- It also offers a comprehensive, all-encompassing pattern of phenomena in real-world contexts.

1.6 CASE STUDIES

Since its introduction, digitalisation is reshaping the solar energy sector through the solar systems in various ways. It is also impacting the solar market and opening new market business. In this research, some examples are taken to gain an insight of the implementation of digital technologies in the solar energy sector as follows.

1.6.1 Sun Exchange and OneWattSolar: Implementing Blockchain Technology to Manage Energy and Facilitate Energy Trading (Digital Technology Used: Blockchain)

The assessment of sustainable energy transitions in sub-Saharan Africa was conducted through a case study of Sun Exchange and OneWattSolar. This may be attributed to the fact that both firms hold prominent positions in the market for enabling the adoption of clean energy by means of digitalisation in sustainable energy transitions. Sun Exchange is a startup company headquartered in South Africa, whereas OneWattSolar is a startup company located in Nigeria. Sun Exchange provides a peer-to-peer marketplace that facilitates the financing and leasing of solar energy system installations. The platform enables anyone worldwide to acquire solar energy-producing cells and generate profits by leasing them to power companies and organisations in developing economies. This scenario establishes a mutually beneficial outcome in which the local communities have access to dependable and affordable electricity, while the investors secure a consistent revenue stream derived from the use fees collected from consumers inside these areas. The model undergoes the following procedure: Sun Exchange collaborates with a consortium of solar energy enterprises, which may be geographically diverse, to find potential projects that might yield substantial benefits through the implementation of small-scale solar installations, namely micro-grids with a capacity of less than one megawatt. As an illustration, these may include healthcare facilities situated in rural regions or communities characterised by an unreliable energy infrastructure. The subsequent phase involves the strategic formulation and cost determination of the installation, with these details being accessible online for potential investors to acquire a portion of the solar cells inside the establishment. Investors have the ability to make purchases in accordance with their own risk tolerance. After the investors have financed the purchase of solar cells, the installation cost is subsequently paid, and the array is created and made operational within a period of 60 days. The local community is subsequently granted the opportunity to utilise the electricity produced by the solar cells, subject to a usage fee structure resembling that of traditional utility payments. In this arrangement, the investors, along with Sun Exchange, receive a portion of the fee commensurate with their respective ownership stakes in the solar array, as per the agreed-upon revenue sharing agreements. Sun Exchange is furthermore accountable for managing the leasing and fee-collecting processes, as well as overseeing insurance and other associated administrative tasks.

OneWattSolar offers a solution that closely resembles the business model employed by Sun Exchange with regards to the funding of solar installations and the distribution of investment returns to the stockholders who provide the financing for these installations. OneWattSolar utilises blockchain technology by including a "smart contract" mechanism to guarantee precise documentation of power use and consumer invoicing, enabling correct evaluation of their usage. In essence, a smart contract refers to a mechanism embedded inside the blocks of a blockchain, designed to digitally assist, verify, or impose the intercession or execution of a contract. These contracts enable the execution of trustworthy transactions that can be traced and are irreversible.

1.6.2 Project Gemini: One of the World's Largest, the U.S. Solar Power Plant Using Digital Intelligent Algorithm and Control System to Optimise and Charge the Battery (Digital Technology Used: Artificial Intelligent)

The Gemini Solar Project is located in Las Vegas, Nevada, and upon completion will be one of the largest solar plants in the United States with 690 MWac/966 MWdc solar arrays and 1,416 MWh of storage capability, and the project will expand over approximately 53 km. The solar facility will consist of several solar array blocks, each containing either standard or bifacial photovoltaic modules. These modules will be installed on horizontal, single-axis trackers. Each block of the array will consist of 32 rows of solar panels, with a spacing of 6 metres between each row. The vertical distance from the bottom of each panel to the ground will exhibit variation ranging from .3 to 2.4 metres. The proposed integrated battery energy storage system would comprise of 425 units of 5 MWh, four-hour battery storage systems. These systems will be utilised to store surplus power generated by the PV panels. The project is being developed by Primergy Solar and NV energy with a Power Purchase Agreement under 25 years with a storage capacity of $1.2 billion. The project will use intelligent algorithms to optimise the charging and discharging commands through the energy management system (EMS). The EMS will regulate the opportunity cost of charging/discharging the battery compared to passing PV-generated energy directly to the grid. The system will also have daily, seasonal, and yearly factors incorporated into its decision-making capability in order to control the supply in accordance with the seasons of the year by comparing them. Gemini's control platform also will include a distributed monitoring and control system at the battery cell level. The battery management system (BMS) ensures appropriate balancing between cells, modules, and racks to prevent premature degradation or overheating. With millions of individual battery cells, the network of BMS controllers uses advanced distributed control techniques to monitor and maintain operation.

1.6.3 Noor Abu Dhabi: The World Leading Power Plant Implementing AI Robots for Cleaning (Digital Technology Used: Artificial Intelligent)

One of the world's largest solar power plants, Noor Abu Dhabi was built in around 20 months. Spread over 8 kilometres square, the number of panels is estimated at 3.2 million. The plant uses 1,400 waterless robots incorporated with AI algorithms for cleaning systems; these travel 800 kilometres twice daily across the solar panels to maximise the performance of the solar panels. The plant's total capacity is 1.2 GW and the cost is estimated AED3.2bn (£695m). It is the world's biggest single-site solar power plant and can produce electricity for 90,000 homes in order to reduce greenhouse gas emissions.

1.7 RESULTS AND DISCUSSION

As a result, all of these companies are achieving great success in the implementation of digital technologies and digital platforms for the usage of the solar energy when it come to the maintenance, management, monitoring and also in the development of new business models. The number of investors and customers is increasing year by year, more solar panels are being installed, and more Gigawatts produced. Sun Exchange and OneWattSolar, by their new business model, are taking the global market in the sub-Saharan Africa to another level and are providing new revenues for consumers and other stakeholders. They are making lots of effort filling the energy gap in several areas in Africa and around the globe.

Gemini Solar Project and Noor Abu Dhabi illustrate the application of AI in the solar energy industry. AI is made of algorithm at its core and these two projects show how suitably this digital technology can be successfully integrated into the solar system through the maintenance robots and the transformation of batteries into smart devices to extend battery lives and provide information for better management of the energy. In the project work the role and importance of digitalisation in the solar energy sector also has its role in the process of transition of solar energy towards sustainable energy resources. The business opportunity that the combination of both offers to all the stakeholders and in in the global market through technologies such as blockchain, AI, IoT, and big data were also studied. Some solutions to limitations of digitalisation in the solar energy sector, such as integration in modern societies and the grid through the case studies of the implementation of digital technologies, as well as more insight on the role and application of the most prominent digital technologies in solar systems, were highlighted. Renewable energy sources, particularly solar energy, are extensively utilised by enterprises and project endeavours for the purpose of generating electricity. The process of digitalisation plays a crucial role in facilitating the attainment of carbon neutrality by leveraging its substantial potential. This potential is harnessed through advancements in data analytics and connectivity. By doing so, digitalisation has the capacity to significantly enhance the overall efficiency of energy infrastructure and reduce costs associated with energy consumption. The recognition and strategic development of these possibilities are of utmost importance for the relevant stakeholders in order to maximise their benefits. The role of government policies, rules, and regulatory frameworks in the digitisation of the energy system is crucial for the development of a secure, sustainable, and intelligent energy future. Based on the analysis of relevant literature and a comprehensive case study, it is evident that the integration of AI, IoT, smart energy grids, and blockchain technology holds promise as a feasible approach to address the challenge of delivering accessible and sustainable energy to marginalised communities worldwide. The energy poverty situation in Africa has resulted in a significant rise in private electricity generation. The research conducted on Sun Exchange and OneWattSolar highlights the potential for Africa's sustainable energy transitions to be digitally enabled through the adoption of digital technologies like smart grids and blockchain technology in the energy sector. This indicates a commercially feasible opportunity for digitalisation in Africa's energy sector. This has the potential to address the issue of energy poverty in Africa and mitigate the shortcomings resulting from the limited accessibility of conventional electricity networks in sub-Saharan Africa. The present level of sustainable energy transitions reveals that digital technologies have the potential to provide a more effective means of overcoming the infrastructure gap necessary to tackle energy poverty in several countries.

Finally, though digitalisation in solar energy provides a huge transformation towards a better future it still has to be studied with more caution before implementation as it come with lots of capital investment, a large amount of power consumption by the system, and safety issues. The implementation of IoT requires the integration of more sensors which end up with more power consumption by the system. Technologies such as blockchain and big data analysis consume more power and require skilled operators. If they are better implemented, especially at large scale, it does make up for the investment, but still deeper assessment needs to be done before implementing those technologies.

1.8 IMPLICATIONS OF THIS RESEARCH

With an emphasis on tackling the major concerns of sustainability, dependability, and customer interaction, legislative and regulatory developments along with technology innovation will probably influence the future of digital transformation in the renewable energy sector. The energy industry can spur growth and create a more just and sustainable energy system in the future by tackling these issues.

1.9 CONCLUSION AND FUTURE RESEARCH OPPORTUNITIES

In conclusion, solar energy integrated with digital technologies takes solar energy to high level among the renewable energy resources in order to achieve more sustainability, more efficiency, more competitiveness, and more reliability as an energy resource. Digitalisation through digital technologies such as AI and IoT helps in managing and monitoring the entire solar system from the generation of power by the producer to the consumption by the end users. By implementing digital technologies, the maintenance and installation of solar systems are made easier and costs of the processes are significantly reduced. Through the new business models that digitalisation provides, blockchain and various digital platforms such as peer to peer unlock new revenue incentives to the PV owners and investors in the global market. Through advanced data analytics using big data these technologies also make solar energy more reliable by providing better forecasting and prediction regarding the amount of energy the solar system could produce on a yearly basis, hence facilitating integration to the grid. All these advantages make the integration of digitalisation in the solar energy sector crucial.

In the future, digital technologies can be used for the study and evaluation of the implementation of digital technologies at reduced cost with less power consumption and also to address some of the more persisting drawbacks in the renewable energy sector such as the effects of its processing and operation on nature and wildlife. For instance, solar farms are burning a large number of birds and wind plants too are causing the death of several birds every year. So why not investigate possible solutions for those drawbacks and other limitations of renewable energy using these high-level digital technologies to minimise risks to human life and encourage effort in those sectors such as intelligent maintenance of solar farms and wind plants by AI drones and robots. Due to the novelty of these technologies in the sector, there are several regulatory issues that still need to be addressed along with the study of the methods to impart digital skills to the relevant operators.

REFERENCES

Abbasi, K.R., Adedoyin, F.F., Abbas, J. and Hussain, K., 2021. The impact of energy depletion and renewable energy on CO_2 emissions in Thailand: Fresh evidence from the novel dynamic ARDL simulation. *Renewable Energy*, *180*, pp. 1439–1450. https://doi.org/10.1016/j.renene.2021.08.078

Alankrita, and Srivastava, S.K., 2020. Application of artificial intelligence in renewable energy. In *2020 international conference on computational performance evaluation (ComPE)* (pp. 327–331). IEEE. https://doi.org/10.1109/ComPE49325.2020.9200065

Azam, W., Khan, I. and Ali, S.A., 2023. Alternative energy and natural resources in determining environmental sustainability: A look at the role of government final consumption expenditures in France. *Environmental Science and Pollution Research*, *30*(1), pp. 1949–1965. https://doi.org/10.1007/s11356-022-22334-z

Baidya, S., Potdar, V., Ray, P.P. and Nandi, C., 2021. Reviewing the opportunities, challenges, and future directions for the digitalization of energy. *Energy Research & Social Science*, *81*, p. 102243. https://doi.org/10.1016/j.erss.2021.102243

Barinova, V., Devyatova, A. and Lomov, D., 2021. The role of digitalization in the global energy transition. *International Organisations Research Journal*, *16*(4), pp. 126–145. https://doi.org/10.17323/1996-7845-2021-04-06

Belu, R., 2013. Artificial intelligence techniques for solar energy and photovoltaic applications. In *Handbook of Research on Solar Energy Systems and Technologies* (pp. 376–436). IGI Global. https://doi.org/10.4018/978-1-4666-1996-8.ch015

Biswal, J. N., Muduli, K., & Satapathy, S. (2017). Critical analysis of drivers and barriers of sustainable supply chain management in Indian thermal sector. *International Journal of Procurement Management*, *10*(4), 411–430. https://doi.org/10.1504/IJPM.2017.085033

Borowski, P.F., 2021. Digitization, digital twins, blockchain, and industry 4.0 as elements of management process in enterprises in the energy sector. *Energies*, *14*(7), p. 1885. https://doi.org/10.3390/en14071885

Chen, W., Alharthi, M., Zhang, J. and Khan, I., 2023. The need for energy efficiency and economic prosperity in a sustainable environment. *Gondwana Research*. https://doi.org/10.1016/j.gr.2023.03.025

Chhotaray, P., Behera, B.C., Moharana, B.R., Muduli, K. and Sephyrin, F.T.R., 2024. Enhancement of Manufacturing Sector Performance with the Application of Industrial Internet of Things (IIoT). In *Smart Technologies for Improved Performance of Manufacturing Systems and Services* (pp. 1–19). CRC Press.

Daly, R., 2022. How can digital help us become more environmentally sustainable? supported by The National Lottery Heritage Fund, licensed under CC BY 4.0. https://www.culturehive.co.uk/digital-heritage-hub/resource/planning/how-can-digital-help-us-become-more-environmentally-sustainable/

Eisenhardt, K.M. 1989. Building theories from case study research. *Academy of Management Review*, *14*(4), pp. 532–550.

Elsaraiti, M. and Merabet, A., 2022. Solar power forecasting using deep learning techniques. *IEEE Access*, *10*, pp. 31692–31698. https://doi.org/10.1109/ACCESS.2022.3160484

Embia, G., Mohamed, A., Moharana, B.R. and Muduli, K., 2023a. Edge Computing-Based Conditional Monitoring. *Intelligent Manufacturing Management Systems: Operational Applications of Evolutionary Digital Technologies in Mechanical and Industrial Engineering* (pp. 249–270). https://doi.org/10.1002/9781119836780.ch10

Embia, G., Moharana, B.R., Mohamed, A., Muduli, K. and Muhammad, N.B., 2023b. 3D printing pathways for sustainable manufacturing. In *New Horizons for Industry 4.0 in Modern Business* (pp. 253–272). Cham: Springer International Publishing https://doi.org/10.1007/978-3-031-20443-2_12

Embia, G. J., Moharana, B. R., Behera, B. C., Mohmaed, N. H., Biswal, D. K., and Muduli, K., 2023. Reliability prediction using machine learning approach. In *Smart Technologies for Improved Performance of Manufacturing Systems and Services* (pp. 21–37). CRC Press.

Felicio, D., 2022. How digital technology enables sustainable energy use, Siemens advanta. https://www.siemens-advanta.com/blog/technology-digitalization-sustainable-future

Galgal, K.N., Ray, M., Moharana, B.R., Behera, B.C. and Muduli, K., 2024. Quality Control in the Era of IoT and Automation in the Context of Developing Nations. In *Smart Technologies for Improved Performance of Manufacturing Systems and Services* (pp. 39–50). CRC Press.

Grimaccia, F., Leva, S. and Niccolai, A., 2017. PV plant digital mapping for modules' defects detection by unmanned aerial vehicles. *IET Renewable Power Generation*, *11*(10), pp. 1221–1228. https://doi.org/10.1049/iet-rpg.2016.1041

Hu, T., Zheng, M., Tan, J., Zhu, L. and Miao, W., 2015. Intelligent photovoltaic monitoring based on solar irradiance big data and wireless sensor networks. *Ad Hoc Networks*, *35*, pp. 127–136. http://doi.org/10.1016/j.adhoc.2015.07.004

Hu, W., Wu, Q., Anvari-Moghaddam, A., Zhao, J., Xu, X., Abulanwar, S.M. and Cao, D., 2022. Applications of artificial intelligence in renewable energy systems. *IET Renewable Power Generation*, *16*(7), pp. 1279–1282. https://doi.org/10.1049/rpg2.12479

Jiang, H., Wang, K., Wang, Y., Gao, M. and Zhang, Y., 2016. Energy big data: A survey. *IEEE Access*, *4*, pp. 3844–3861.

Juszczyk, O. and Shahzad, K., 2022. Blockchain technology for renewable energy: Principles, applications and prospects. *Energies*, *15*(13), p. 4603. https://doi.org/10.3390/en15134603

Kalogirou, S. and Sencan, A., 2010. Artificial intelligence techniques in solar energy applications. *Solar Collectors and Panels, Theory and Applications*, *15*, pp. 315–340.

Kangas, H.L., Ollikka, K., Ahola, J. and Kim, Y., 2021. Digitalisation in wind and solar power technologies. *Renewable and Sustainable Energy Reviews*, *150*, p. 111356. https://doi.org/10.1016/j.rser.2021.111356

Kumar, N.M., Atluri, K. and Palaparthi, S., 2018, March. Internet of Things (IoT) in photovoltaic systems. In *2018 national power engineering conference (NPEC)* (pp. 1–4). IEEE. https://doi.org/10.1109/NPEC.2018.8476807

Malik, H., Chaudhary, G. and Srivastava, S., 2022. Digital transformation through advances in artificial intelligence and machine learning. *Journal of Intelligent & Fuzzy Systems*, *42*(2), pp. 615–622. https://doi.org/10.3233/JIFS-189787

Mangla, S.K., Luthra, S., Jakhar, S., Gandhi, S., Muduli, K. and Kumar, A., 2020. A step to clean energy-Sustainability in energy system management in an emerging economy context. *Journal of Cleaner Production*, *242*, p. 118462. https://doi.org/10.1016/j.jclepro.2019.118462

Motlagh, N.H., Mohammadrezaei, M., Hunt, J. and Zakeri, B., 2020. Internet of Things (IoT) and the energy sector. *Energies*, *13*(2), p. 494. https://doi.org/10.3390/en13020494

Muduli, K., Barve, A., Tripathy, S. and Biswal, J.N., 2016. Green practices adopted by the mining supply chains in India: A case study. *International Journal of Environment and Sustainable Development*, *15*(2), pp. 159–182. https://doi.org/10.1504/IJESD.2016.076365

Muduli, K., Kusi-Sarpong, S., Yadav, D.K., Gupta, H. and Jabbour, C.J.C., 2021. An original assessment of the influence of soft dimensions on implementation of sustainability practices: Implications for the thermal energy sector in fast growing economies. *Operations Management Research*, *14*, pp. 337–358. https://doi.org/10.1007/s12063-021-00215-x

Nexus-Integra, 2023. Digital transformation in the renewable energy sector. https://nexusintegra.io/digital-transformation-renewable-energy/

Nwaiwu, F., 2021. Digitalisation and sustainable energy transitions in Africa: Assessing the impact of policy and regulatory environments on the energy sector in Nigeria and South Africa. *Energy, Sustainability and Society*, *11*(1), pp. 1–16. https://doi.org/10.1186/s13705-021-00325-1

Pagliaro, M. and Meneguzzo, F., 2019. Digital management of solar energy en route to energy self-sufficiency. *Global Challenges*, *3*(8), p. 1800105. https://doi.org/10.1002/gch2.201800105

Pakulska, T. and Poniatowska-Jaksch, M., 2022. Digitalization in the renewable energy sector—New market players. *Energies*, *15*(13), p. 4714. https://doi.org/10.3390/en15134714

Piso, K., Mohamed, A., Moharana, B.R., Muduli, K. and Muhammad, N., 2023. Sustainable Manufacturing Practices through Additive Manufacturing: A Case Study on a Can-Making Manufacturer. *Intelligent Manufacturing Management Systems: Operational Applications of Evolutionary Digital Technologies in Mechanical and Industrial Engineering*, pp. 349–375. https://doi.org/10.1002/9781119836780.ch14

Quixy Editorial Team, 2023. How digital transformation in solar industry can make it shine. https://quixy.com/blog/digital-transformation-in-solar-energy-industry/

Ramjeawon, T., 2020. *Introduction to Sustainability for Engineers*. CRC Press.

Sahoo, S.K., 2016. Renewable and sustainable energy reviews solar photovoltaic energy progress in India: A review. *Renewable and Sustainable Energy Reviews*, *59*, pp. 927–939. https://doi.org/10.1016/j.rser.2016.01.049

Sharifi, A., Ahmadi, M. and Ala, A., 2021. The impact of artificial intelligence and digital style on industry and energy post-COVID-19 pandemic. *Environmental Science and Pollution Research*, *28*, pp. 46964–46984. https://doi.org/10.1007/s11356-021-15292-5

Singh, R., Akram, S.V., Gehlot, A., Buddhi, D., Priyadarshi, N. and Twala, B., 2022. Energy System 4.0: Digitalization of the energy sector with inclination towards sustainability. *Sensors*, *22*(17), p. 6619. https://doi.org/10.3390/s22176619

Spanias, A.S., 2017. Solar energy management as an Internet of Things (IoT) application. In *2017 8th International Conference on Information, Intelligence, Systems & Applications (IISA)* (pp.1–4). IEEE. https://doi.org/10.1109/IISA.2017.8316460

Sribna, Y., Koval, V., Olczak, P., Bizonych, D., Matuszewska, D. and Shtyrov, O., 2021. Forecasting solar generation in energy systems to accelerate the implementation of sustainable economic development. *Polityka Energetyczna*, *24*. https://doi.org/10.33223/epj/141095

Taşaltin, N., 2019. Digitalization of solar energy: A perspective. *Journal of Scientific Perspectives*, *3*(1), pp. 41–46. https://doi.org/10.26900/jsp.3.005

Themistocleous, M., Stefanou, K. and Iosif, E., 2018. Blockchain in solar energy. *Cyprus Review*, *30*(2), pp. 203–212.

Torres, J.F., Troncoso, A., Koprinska, I., Wang, Z. and Martínez-Álvarez, F., 2019a. Deep learning for big data time series forecasting applied to solar power. In *International Joint Conference SOCO'18-CISIS'18-ICEUTE'18: San Sebastián, Spain, June 6–8, 2018 Proceedings 13* (pp. 123–133). Springer International Publishing. https://doi.org/10.1007/978-3-319-94120-2_12

Torres, J.F., Troncoso, A., Koprinska, I., Wang, Z. and Martínez-Álvarez, F., 2019b. Big data solar power forecasting based on deep learning and multiple data sources. *Expert Systems*, *36*(4), p. e12394. https://doi.org/10.1111/exsy.12394

Truong, T.C., 2022. The impact of digital transformation on environmental sustainability. *Advances in Multimedia*, *2022*, pp. 1–12. https://doi.org/10.1155/2022/6324325

Valeeva, Y., Kalinina, M., Sargu, L., Kulachinskaya, A. and Ilyashenko, S., 2022. Energy sector enterprises in digitalization program: Its implication for open innovation. *Journal of Open Innovation: Technology, Market, and Complexity*, *8*(2), p. 81. https://doi.org/10.3390/joitmc8020081

Wang, J., Ma, X., Zhang, J. and Zhao, X., 2022. Impacts of digital technology on energy sustainability: China case study. *Applied Energy*, *323*, p. 119329. https://doi.org/10.1016/j.apenergy.2022.119329

Watson, B., 2017. Here comes the sun: This digital technology will make solar power more predictable, Invenergy.https://www.ge.com/news/reports/comes-sun-digital-technology-will-make-solar-power-predictable

Wörner, A., Meeuw, A., Ableitner, L., Wortmann, F., Schopfer, S. and Tiefenbeck, V., 2019. Trading solar energy within the neighborhood: Field implementation of a blockchain-based electricity market. *Energy Informatics*, *2*, pp. 1–12. https://doi.org/10.1186/s42162-019-0092-0

Wu, Y.K., Lai, Y.H., Huang, C.L., Phuong, N.T.B. and Tan, W.S., 2022. Artificial Intelligence Applications in estimating invisible solar power generation. *Energies*, *15*(4), p. 1312. https://doi.org/10.3390/en15041312

Xie, L., Huang, T., Zheng, X., Liu, Y., Wang, M., Vittal, V., Kumar, P.R., Shakkottai, S. and Cui, Y., 2022. Energy system digitization in the era of AI: A three-layered approach toward carbon neutrality. *Patterns*, *3*(12). https://doi.org/10.1016/j.patter.2022.100640

Yousif, J., 2021. Implementation of Big Data analytics for simulating, predicting & optimizing the solar energy production. *The Journal of Computational and Applied*, *1*, pp. 133–140. https://doi.org/10.52098/acj.202140

Zhang, H., Gao, S. and Zhou, P., 2023. Role of digitalization in energy storage technological innovation: Evidence from China. *Renewable and Sustainable Energy Reviews*, *171*, p. 113014. https://doi.org/10.1016/j.rser.2022.113014

2 Green Energy
A Sustainable Approach for Water Desalination Using Solar Still

Dillip Kumar Biswal
Krupajal Engineering College, Bhubaneswar, India

Bikash Ranjan Moharana and Kamalakanta Muduli
Papua New Guinea University of Technology, Lae, Papua New Guinea

FonoTamo Romeo Sephyrin
University of Tennessee, Knoxville, Tennessee, U.S.A.

2.1 INTRODUCTION

The Earth's surface is predominantly covered by water, but a relatively small fraction of that is freshwater, and even less of that is readily available to the billions of living beings that depend on freshwater for survival. Additionally, the demand for drinking water is continuously expanding as a result of the increase in the global population.

Presently, the most significant challenge lies in bridging the disparity between the need for and the availability of clean potable water for every living organism on Earth. Contaminated water typically contains harmful elements such as bacteria, viruses, parasites, undisclosed substances, and both chemical and physical pollutants. Consuming such water poses severe health risks. As a result, it is essential to investigate alternate approaches and technologies for purifying contaminated or unclean water from various sources such as rainwater, wells, rivers, lakes, and oceans. This would enable the cost-effective production of safe drinking water.

Despite the availability of multiple water purification technologies, it's important to note that these methods rely on conventional energy sources, primarily electricity, and involve various components such as batteries, filters, or membranes. In contrast, solar distillation stands out as a straightforward, cost-efficient, and environmentally friendly approach to water purification.

DOI: 10.1201/9781003349877-2

Figure 2.1(a) illustrates different categories of solar desalination methods for clarity. Solar desalination stands as a widely adopted treatment solution on a global scale. This technique utilizes solar energy, a sustainable power source, to generate clean and drinkable water. As illustrated by Figure 2.1(b), there are two categories of "solar stills": active and passive.

In their historical overview on water desalination, Nebbia and Menozzi (1966) offer an extensive historical context for solar distillation. On the other hand, Garg and Mann (1976) take a different approach by examining the evolution over time of solar distillation technologies, delving into research focused on passive solar distillation methods.

Layout of a passive "solar still" system is presented in Figure 2.2, where meticulous insulation with silicon sealant is applied to all enclosure components, except for the glass cover, to minimize heat loss. While in operation, the transparent glass cover permits sunlight to penetrate through, which subsequently directly heats the absorbing plate. From the absorption plate, saline water vapour flows and gathers on the inner surface of the glass plate. The passage through which saline water is introduced at the top of the absorption and glass plates allows for evaporation of some of the water, with the remaining liquid being gathered at the bottom and then removed as concentrated brine.

2.2 RESEARCH ON PASSIVE SOLAR DESALINATION SYSTEMS

Researchers have explored two primary categories of solar distillation systems: active and passive. In a solar still, distillate production is solely regulated by solar energy. In active solar stills, the incorporation of a receiver and solar power contributes to the improvement in distillate output. Next, we will outline the research effort dedicated to a passive solar still.

Moustafa and Brusewitz (1979) conducted research to enhance distillate production, and they achieved this by designing and building a solar powered device with a wicking composition and a water flow mechanism controlled by a flow regulator and a shut-off valve. Their findings indicated that employing the water flow system yielded more favourable results in increasing distillate output. Prakash and Kavathekar (1986) demonstrated the effectiveness of a solar still with a regenerative design, harnessing solar energy, in conditions differing from the typical environment within the same geographic area and climate. The findings have been contrasted to those of a conventional solar still. A transient analytic solution for single-basin solar still with internal water flow has been presented (Yadav and Kumar, 1991). The study extensively investigates the system's performance, taking into account various factors. A comparison was made between the yield of upward-oriented, double-effect and closed-type solar stills, both of which were positioned on a flat surface at a 10° angle and found that due to the more significant rise in water temperature in upward-oriented double-effect solar stills, these systems yielded more efficient results compared to closed-type solar stills (Yeh, 1993). Mowla and Karimi (1995) have created a mathematical model that calculates the distillate yield from seawater, taking into account various climatic conditions and specifications of the solar still. Their findings revealed a high degree of conformity between the estimated results of the model

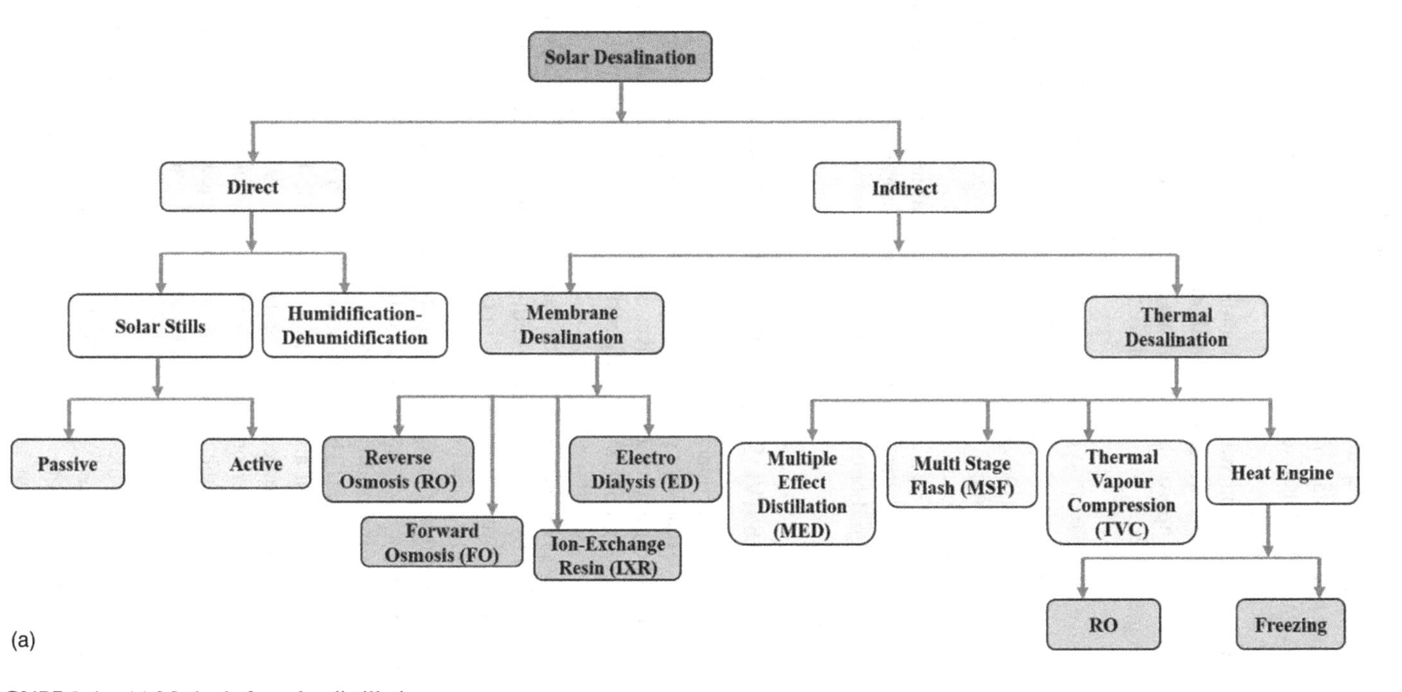

FIGURE 2.1 (a) Methods for solar distillation.

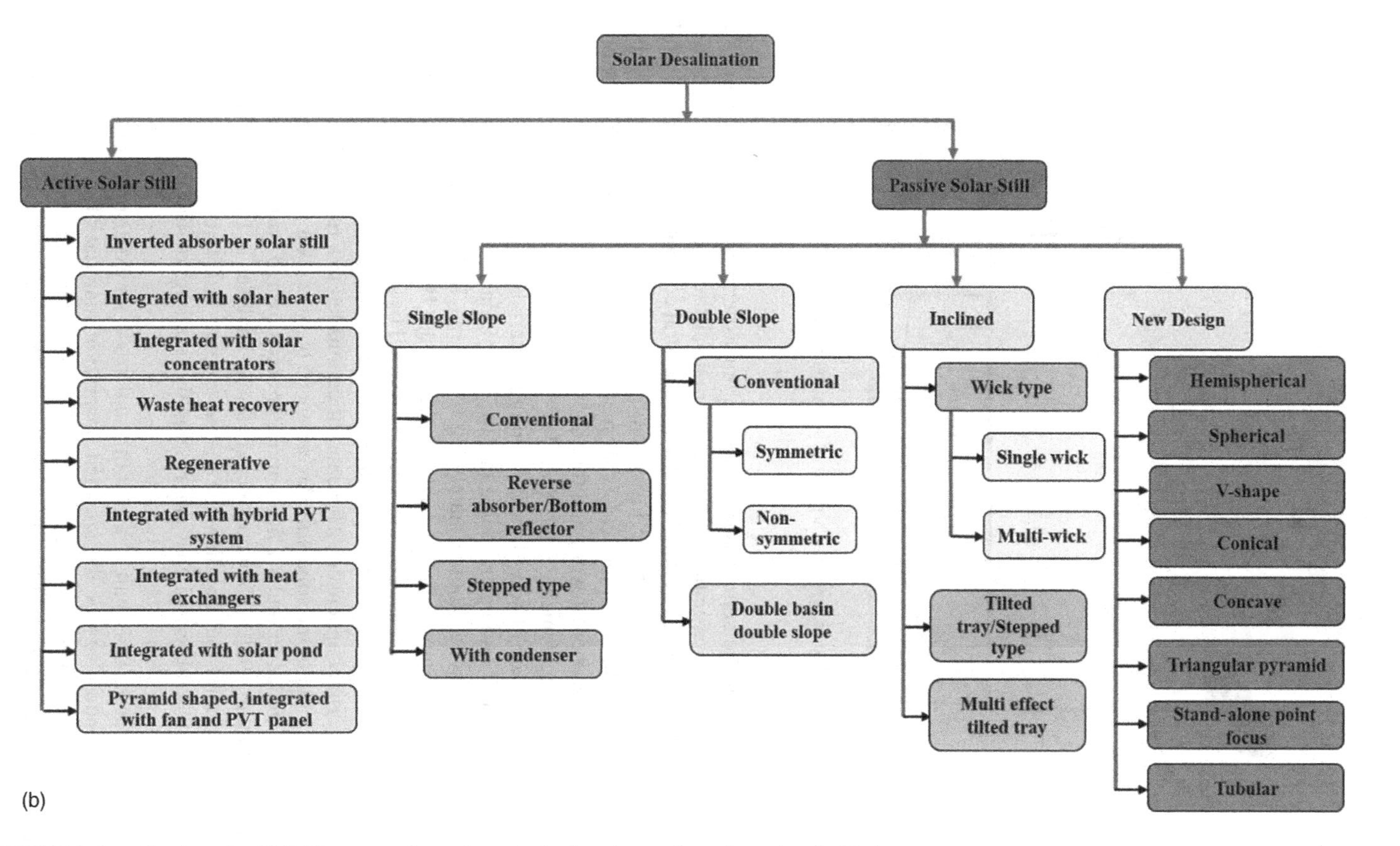

FIGURE 2.1 (Continued) (b) Different configurations for both active and passive solar distillation systems.

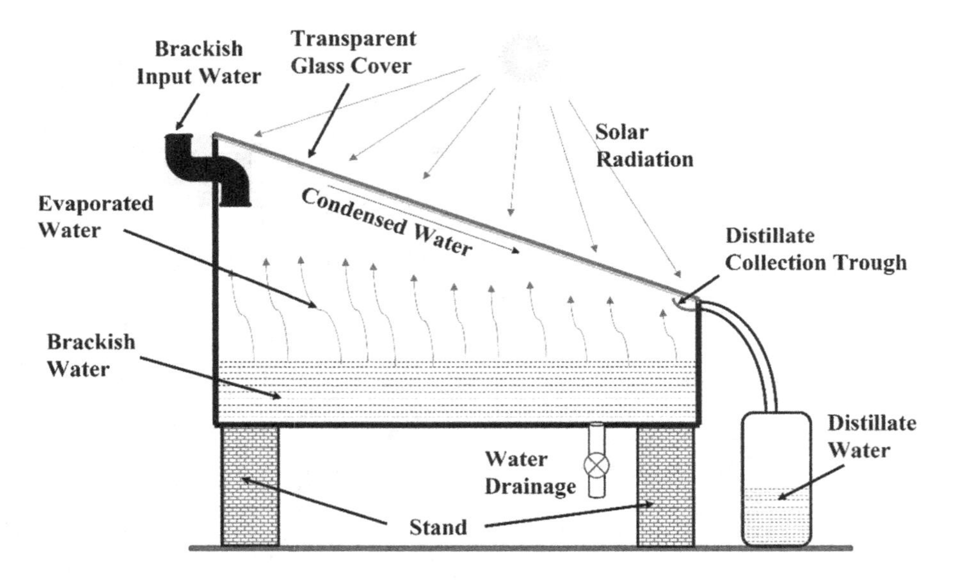

FIGURE 2.2 Schematic diagram of a passive solar distiller.

they developed and the actual data obtained from experiments. Aboul-Enein et al. (1998) have devised a basic momentary mathematical framework that relies on analytically solving the energy balance equation for several components of the solar still. They conducted a comprehensive investigation of the still's thermal performance, both in theory and through practical experiments. Remarkably, the results obtained from both the experimental work and the theoretical model exhibited a high degree of consistency and agreement. Bilal et al. (1998) carried out an investigation in order to estimate the efficacy of several heat-retaining materials, in the context of an inactive solar still. Their findings revealed that the use of black dye led to a remarkable increase in productivity, approximately 60%, making it the most effective heat-absorbing material among those tested. Khalifa et al. (1999) achieved a significant enhancement in distillate production from a solar still by introducing several modifications. They accomplished this by heating the saline water and incorporating both internal and external condensers. These modifications led to a remarkable escalation in the distillate yield of the solar still. An enhanced passive solar still design featuring a flat opening condenser and a 1 m^2 evaporator area covered with 6 mm–thick glass (El-Bhai and Inan, 1999). They conducted various analyses at Hacettepe University in Turkey and found that the distillate output aligned well with previous research. Furthermore, they observed that improving the cooling of the condenser cover boosted efficiency from 48% to over 70%.

In their research, El-Sebaii et al. (2000) undertook an experiment to assess how the thermal conductivity of the attached absorption material affects the everyday performance of a standard passive solar still. They tested various materials, including mica, copper, aluminum, and stainless-steel plates as suspended absorbers. The findings indicated that the solar still modified with mica exhibited a 42% higher enhancement in daily yield than the standard single-basin solar still. Their conclusion was that suspended plates proved to be more efficient when dealing with larger quantities of basin water. Djebdedjian and Rayan (2000) utilized the Navier-Stokes equation to analyze the functioning of a standard solar still. They established a scientific model that incorporated a combination of air and water vapour and found that it closely matched their experimental results. Al-Hinani et al. (2002) did a mathematical study of a traditional solar still to explore the most suitable design factors for the angle of tilt of the glass cover, the influence of insulation, and the presence of a black coating on the safety plate.

Voropoulos et al. (2002) conducted a comprehensive analysis of a solar still performance, considering the specific operating conditions and atmospheric data relevant to Greece. Their research suggested that the distillate yield of the conventional solar still was substantially influenced by the prevailing operating conditions and essential atmospheric parameters. The effectiveness and validity of their approach were affirmed by the alignment observed between the anticipated and actual water output, as well as between the experimental results and the theoretically calculated characteristic parameters. Fath Hassan and Hosny (2002) encompassed an examination of the impacts of different factors related to design, operations, and environmental conditions. Their findings demonstrated that the incorporation of balancing on the front side of a passive solar still led to a remarkable 55% increase in the distillate yield. Valsaraj (2002) conducted a study focusing on a solar still design featuring a single slope and a single basin. In this design, a perforated and folded aluminum sheet is floated on the saline solution's surface to effectively concentrate solar radiation. This approach, particularly when there is a substantial water depth, prevents the cooling of the entire water mass and consequently results in a significant boost in distillate production. Bassam and Rababa'h (2003) explored the enhancement of surface area in passive solar stills by employing sponge cubes of varying sizes. Their research involved multiple trials with sponge cubes of consistent dimensions placed at different depths within the still. The outcome of their experiments revealed a significant improvement in distillate production, with the capillary effect of water inside the sponge cubes leading to a remarkable increase in yield, ranging from 18% to 273% when contrasted with an equivalent still running in the same way but without sponge cubes. Abdallah and Badran (2003) introduced an innovative concept wherein they proposed the utilization of a computerized sun-tracking mechanism to enhance the distillate output of a passive solar still. Their research objectives centered on tracking the sun's movement to maximize distillate production, and their investigation yielded remarkable results in terms of augmenting the generation of potable water from solar stills. They ultimately concluded that the sun-tracking system outperforms a stationary setup and is, overall, a more effective and efficient approach. Naim and Mervat (2003) have created an innovative single-stage solar still that advances energy efficiency by incorporating a phase change energy storage mixture. The concentration and flow rate of saline water, the material of choice for energy storage, and the temperature of the entering water were all carefully examined in order to determine how they affected the still's output. It's interesting to note that the distilled water produced as a result of using an energy storage material was improved. However, it's worth noting that as the concentration of saline water increased, the efficiency of distilled yield actually decreased.

For a conventional passive solar still, Shukla (2003) created a computer-based model to evaluate the possible distillate yield in relation to the inner glass temperature. Hanson et al. (2004) investigated the application of a "single-basin solar still" to eliminate a particular range of inorganic, bacterial, and organic pollutants through laboratory and field research. Shukla and Sorayan (2005) devised a method for improving the final output of a solar still by employing jute fibres. Their research revealed that using jute fabric could enhance evaporation by reducing the water's salinity in the basin. Furthermore, they conducted both theoretical and practical experiments, leading to a comprehensive mutual understanding of their findings. In order to increase distillate production, Zeinab and Ashraf (2007) investigated a single-slope solar still using a variety of solar-absorbing materials such glass, rubber, and black stone. They investigated different materials for six months and found that black stone, rubber, and glass were the most successful at boosting the yield of the passive solar still. According to their findings, the modification boosted freshwater productivity on an average of 18%. Nijmeh et al. (2005) investigated the impact of several sunlight-based energy absorption materials on the distillate production of a detachable solar still in Jordan's atmospheric conditions. They modified the procedure by adding dissolved salts, violet dye, and charcoal. By utilizing potassium permanganate in a still powered by sunlight, they were able to improve distillate production by 26%. Omri et al. (2005) developed numerical modeling for natural convection in a triangular cavity with homogeneous radiation from the sun employing the control volume finite element method. Their study showed that the most crucial factors influencing cavity and Rayleigh Number were the flow regime and heat transmission. Kauzo Murase et al. (2006) introduced a novel concept for a passive solar still combined with a water distribution system in Algeria. They conducted both numerical and experimental tests on a tube-type solar still and achieved highly favorable results, demonstrating good agreement between their calculations and practical observations. Ayber (2006) considered slanted wick composition solar still and concluded that the day-to-day output of such solar still is 2.5–3.5 kg/m^2/day for summertime temperature. Apart from purified water, the normal temperature of the water obtained from such a solar still is about 40°C, which is appropriate for domestic uses.

Tanaka and Nakatake (2006) introduced an innovative approach for a latent solar still, which involved the use of both inner and outer reflectors to redirect solar rays towards the condensing bowl. Through various experiments conducted in different atmospheric conditions in China, they found that employing these inner and outer reflectors resulted in a significant 48% increase in distillate production. Tiwari and Tiwari (2006) investigated the efficacy of a passive solar still in New Delhi, India, during summer by altering the depth of brackish water. They detected five distinct water depths at 24-hour intervals over the period of seven days, ranging from 0.04 to 0.18 meters. Their conclusion was that because of the loss in volumetric heat capacity, a lower water depth resulted in a higher distillate production. Omar Badran (2007) demonstrated a solar still execution by varying aspects on distillate output and conducted a few trials in Jordan's environment and projected that when combined with a black-top bowl liner and sprinkler, it still increases distillate production by

51% and increases night-time generation by 16% by combining the aforesaid characteristics. Kumar and Bai (2008) conducted a study on a passive solar still, implementing an enhanced condensation technique aimed at improving the condensation on the inner glass cover using various brine samples. They found that a higher distillate production was achieved when using tap water, as compared to seawater and effluent from the dairy industry, showcasing the effectiveness of their new condensation approach. Torchia-Núñez et al. (2008) performed a long-term transient theoretical analysis of a passive solar still to investigate the factors influencing energy dissipation. They found that ambient temperature had no discernible impact on energy efficiency and recommended that the insulation thickness should exceed 0.02 meters to achieve greater energy productivity. Their analysis also indicated that improved thermodynamic performance was attainable by reducing temperature differentials. Velmurugan and colleagues (2008) conducted several studies on enhancing the distillate production of a stepped solar still through the incorporation of wick cubes and baffles. Their primary focus in the current research was to upsurge the surface area of the saltwater. They achieved a 30% increase in distillate yield through the combined effects of wick cubes and baffles.

Shakthivel and Shanmugasundaram (2008) investigated the use of various sized rocks as solar energy–absorbing materials in a passive solar still. In their research, they used rocks with diameters of 2, 4, and 6 mm. They also conducted a thermal analysis of a solar still using pebbles of various sizes and compared the results to experimental data, observing considerable improvements. Following a thorough investigation, they determined that the passive solar still with 6 mm rocks was the most successful in improving distillate production. Sahoo et al. (2008) aimed to eliminate fluoride from drinking water by employing a solar still while also modifying the design of the bowl liner and incorporating thermocol insulation to enhance efficiency and distillate output. They predicted that with the implementation of an appropriately darkened bowl liner and thermocol insulation, fluoride removal could be achieved at a rate of 92–96%, and productivity could be increased by 6%.

Nafey et al. (2008) performed a test involving a single-basin passive solar still to investigate the impact of surfactant concentration on distillate production. They introduced various surfactant concentrations into the solar still and established that distillate production increased. They came to the conclusion that a 400-ppm injection caused a 6% drop in distillate output. Jiang et al. (2009) constructed a passive solar still desalination system that was integrated with flash equipment. The project's goal aimed to boost the output of distillate by flickering brine in a solar still. To assess agreement, they conducted a theoretical analysis and compared it to an experiment analysis; they observed good agreement. Kabeel (2009) developed and examined a solar still with a concave configuration and a glass cover shaped like a pyramid. They utilized Jute fabric as a wicking material at the base of the solar still to enhance sunlight absorption and boost distillate production through capillary action. Their research revealed a 30% enhancement in the daily efficiency of the solar still when matched to a traditional design.

Aadallah et al. (2009) examined the impact of several wick materials on distillate production in a passive solar still. These wick materials are distributed within the still to enhance the distillation process. Their research encompassed multiple experiments conducted in the environmental conditions of Jordan. They determined that the best wick material was volcanic sand, which increased distillate production by 45%. Feilizadeha et al. (2010) introduced an innovative strategy for evaluating the productivity of a passive solar still, known as a radiation model. They examined the diverse effects of this radiation model on the water surface, side walls, and back walls of the passive solar still. Their conclusion was that the side walls and back walls had a substantial impact upon the distillation yield and effectiveness of the solar still. In the adverse climate of New Delhi, India, they conducted a study on the life cycle cost analysis of single- and double-slope latent solar stills (Dwivedi and Tiwari, 2006). Their research aimed to compare the performance of single- and double-slope solar stills using the same saline solution depth in New Delhi's climate. Their findings led them to conclude that the single-slope latent solar still was more cost effective compared to the double-slope passive solar still.

In context of passive solar still, Kalidasa Murugavel et al. (2010) experimented with several sensible heat storage materials, such as quartzite rock, red brick pieces, cement, concrete blocks, washed stones, and iron pieces. Their findings revealed that quartzite rock proved to be the most effective sensible heat storage material for enhancing the distillate production of the passive solar still. Khaled (2010) designed a latent solar still incorporating a compacted medium to enhance distillate production. They employed a helical copper spring as a flexible packed medium to generate harmonic oscillations. Their research demonstrated that the copper spring had a substantial vibrating effect, resulting in an increase in distillate yield and improvement in efficiency compared to a standard solar still. Kalidasa Murugavel and Srithar (2011) conducted an experiment involving a double-slope passive solar still with minimal water mass in the basin and explored different combinations of energy-absorbing materials and varying fin configurations. Their research revealed that covering the solar still with lightweight cotton cloth and arranging the fins lengthwise led to an enhanced distillate output. In a performance examination of a single-slope solar still Mahdi et al. (2011) looked into the effects of 4 mm plexiglass on the internal heat transfer coefficients of a passive solar still. Additionally, they assessed the performance of a wick solar still in Iran's climate using charcoal as the evaporator material. Their studies comprised varying the brine's depth, flow rate, and salinity. Their findings showed that employing charcoal as the evaporator material was the most effective method for increasing distillate production, especially with a 2 mm depth, lower mass flow rate, and lower brine salinity. When compared to a typical still made of red bricks coated with cement, Kabeel et al. (2020) increased distillate production by 45%.

A modified version with a darkened surface was constructed and installed by Biswal (2021) to improve the efficiency of a single-slope solar still. When operated with water inputs of 20 and 30 liters, respectively, this upgraded solar still achieved efficiencies of 7.41% and 7.95%. Furthermore, comprehensive testing verified that the distilled water passed acceptable criteria. Adamu et al. (2022) discovered that using pebbles as heat storing materials in a solar still had an intriguing outcome. They discovered that the capacity of the basin to store heat increased as the water volume increased, resulting in higher distillate output during periods of reduced sunlight. When the major goal was to optimize distillate output during the day, it was discovered that maintaining less water in the basin was optimum. Khanmohammadi and Khanjani (2021) enhanced the yield of solar still through the application of hydrophobic condensation surfaces and cold plasma technology. According to their projections, a system with a plasma coating produced 25.7% more drinking water than a system without a coating. Mirmanto et al. (2021) studied the possibility of continuously feeding a conventional single-slope solar still with salt water. Three types of absorbers were evaluated experimentally in this paper. The findings of the absorber with 15 fins were found to be superior to those of the flat absorber. Panchal et al. (2021) investigated the solar still efficacy using discarded magnesia bricks as energy storage materials. Kumar et al. (2021) improved desalination performance by employing aluminum fins as solar absorbent material. Mevada et al. (2022) improved the performance by using black granite and marble stones, achieving a daily effectiveness of 72.6% higher than that of a typical solar still. Chen et al. (2022) highlighted the environmental solicitations of carbon compounds produced from plastic trash for long-term green energy. Kaviti et al. (2023b) in their study achieved a 36.35% enhancement in distillate production by incorporating camphor-treated banana stems. Kaviti et al. (2023a) prepared a hybrid nanofluid using a two-step method with cerium oxide nanoparticles and multi-walled carbon nanotubes (MWCNTs) in an 80:20 ratio and discovered that the rate of distilled water production reduces as the content of hybrid nanofluid, that is, cerium oxide and MWCNTs, increases.

Abed et al. (2022) conducted a study where they mathematically modelled a solar desalination system and carried out experimental investigations in Iraq. In these experiments, they introduced thymol blue and orange methyl into the saline water to compare the distillation efficiency with undyed saline water. The results revealed that the addition of these chemical dyes increased the production and efficiency of the solar still. Vigneswaran et al. (2023) aimed to increase the distilled water production of a passive solar still (PSS) by expanding the available condensation surface area. They achieved this by integrating a water jacket along the sidewalls of the PSS.

2.3 EXPLORATION ON DATA-DRIVEN APPROACHES IN A SOLAR DESALINATION SYSTEM

Desalination, a process aimed at removing salt and impurities from water, is commonly achieved through three primary methods: thermal, pressure, and electrical approaches. The oldest of these methods is thermal distillation, where water with a high salt content is heated to the point of boiling, and the resulting steam is condensed and collected. Intelligent methods, such as artificial neural networks (ANNs), are often favoured for their superior accuracy. This is primarily because of their intricate structures, which allow them to represent complex operations with a greater precision. Moreover, research has revealed that combining optimization techniques with intelligent methods enhances accuracy, as it ensures that the factors influencing precision are optimized to their best possible values (Moharana and Sahoo, 2014). The approach and algorithm employed are two of the most essential aspects determining the accuracy of data-driven strategies in estimating solar still outputs (Mashaly and Alazba, 2019).

In their study on predicting the working efficiency of an inclined stepping solar still, Abujazar et al. (2018) considered a wide range of variables. Their approach involved the utilization of cascaded forward ANNs with varying numbers of neurons, alongside a linear model and regression analysis. Zarei and Behyad (2019) applied ANN to create a model for the output of a humidification-dehumidification desalination with solar energy. The apparatus served the dual purpose of moisturizing the interior of a greenhouse and supplying a reservoir of clean water. Forecasting the production of an active solar still, Essa et al. (2020) compared the effectiveness of ANN utilizing the Harris Hawk optimizer to that of regular ANN and support vector machines. As inputs to their models, they used ambient temperature, time, wind speed, sun irradiation, and vapour velocity. Bahiraei et al. (2020) modelled the efficiency of a nanofluidic solar still integrated with a thermoelectric module using ANN in conjunction with genetic algorithm and imperialist competition method. PSO-ANFIS and PSO-ANN were used by Bahiraei et al. (2021) to predict the performance of a solar still using Cu_2O nanoparticles. Benghanem et al. (2021) designed and experimentally verified a prototype of an intelligent solar still for water desalination in Saudi Arabia. Their design significantly accelerated the evaporation process of saline water, leading to a notable enhancement in daily water production. Faegh et al. (2021) used various ANN methodologies to simulate the output ratio and heat transfer rate of a heat pump–assisted desalination system's evaporator and evaporative condenser. Their findings indicated that the R-squared values for all the model outputs exceeded 0.91. Wang et al. (2021) employed three different methods, namely random forest (RF), ANN, and multilinear regression, for the purpose of forecasting the system's productivity. Their results indicated that employing the random forest model yielded the most accurate predictions, exhibiting the smallest degree of deviation when compared to the other techniques. Bagheri et al. (2021) employed ANN to construct a model for a solar desalination system that consisted of various components such as photovoltaic panels, a heater, a battery, and a cylindrical parabolic collector, among others. In a research study conducted by Sohani et al. (2022), various ANN architectures, such as backpropagation, eedforward, and radial basis function, were employed to predict both the water temperature and hourly water output of a solar still featuring an improved design. Kandeal et al. (2021) simulated the performance of a double slope solar still using the carbon black Nano-fluid using linear support vector regression, ANN and random forest.

Salem et al. (2022) conducted optimization on an ANN, specifically a multilayer perceptron, to forecast and identify the efficiency of a solar thermal energy–dependent water desalination system. Additionally, when estimating the output of a modified SS-HDH (solar still-humidification dehumidification), their proposed OMLP model was compared with other intelligence-based models, including the conventional decision tree, multilayer perceptron, and Bayesian ridge regression algorithms. Musharavati et al. (2022) have introduced a novel energy system based on biomass, aimed at generating both power and desalinated water. To accomplish the most effective outcomes in terms of exergy efficiency and overall cost rate, they employed multi-objective grey wolf optimizer algorithm for multiple objective optimization. The study also delved into the connection among the objective function and decision variables, utilizing ANN to identify the energy system's ideal operating point.

Zouli (2023) presents an innovative and highly efficient method that leverages neural networks for forecasting a hybrid solar-powered desalination system's performance. The proposed approach recommends employing reptile search optimization in conjunction with radial basis long short-term memory to envisage the effectiveness of the hybrid solar-powered desalination process.

Khanmohammadi et al. (2023) introduce a multi-generation solar system that includes a humidification-dehumidification desalination unit, super- and trans-critical CO_2 cycles, and solar heliostat fields. They have undertaken mathematical modelling using Engineering Equation Solver and conducted a multi-criteria optimization of this proposed system using MATLAB software. Mohapatra et al. (2022) have undertaken a study that explores the application of data-driven machine learning models for the analysis and forecasting the design characteristics and performance of evacuated U-tube solar collectors. Neel Shrimali et al. (2023) employed Bayesian optimization to fine-tune hyper parameters during the model training process, achieving an autoregressive approach. They utilized the Gaussian process regression technique to construct the predictive model.

Saba et al. (2023) adopted a Monte Carlo approach to evaluate how uncertainties related to inflation, discount rates, and emission factors affect the overall technical, environmental, and economic feasibility of the system under consideration. They focused on two critical indicators, namely the cost per liter and the annual carbon dioxide savings, as essential metrics for assessing the system. Their outcomes indicated that the most suitable probability distribution functions for modeling these uncertainties were the normal distribution for inflation, the log-normal distribution for discount rates, and a combination of the two for emission factors.

2.4 CONCLUSION

Clean water is a vital resource for human survival, but the fast-growing pollution and population have led to the contamination of water sources. Earth has limited sources of freshwater, and there's an imperative need to secure additional freshwater for human sustenance and everyday needs. Desalination has emerged as the most effective solution to access freshwater since about 70% of the earth's surface is enveloped by saline water. Utilizing solar stills is an efficient method to maximize the benefits of desalination and generate a greater volume of freshwater. Ongoing research efforts are exploring various aspects of solar still technology to enhance its effectiveness in drinkable water production. This research provides a current summary of studies related to solar stills, encompassing both experimental and computational investigations. Leveraging computational studies can streamline the time-consuming experimental processes. This comprehensive review serves as a valuable resource for scientists, industry professionals, and researchers engaged in this field. The concept of solar still could also solve the rural clean drinking water problem of the emerging economics, which leads to achieve SDG#10, that is., "reduced inequalities". It also helps to achieve SDG#3, "good health and well-being", SGD#6, "clean water and sanitation" and SDG#7, "affordable and clean energy". SDG#3 seeks to provide equitable access to healthcare services as well as universal health coverage. SDG#6 ensures accessibility of drinking water to everyone with sustainable management and cleanliness as well. Since the solar still relies predominantly on solar energy, it satisfies the SDG#6, which aim to make sure everyone has access to modern, affordable, dependable, and sustainable energy. The following are the key takeaways that can be drawn from this review:

- Desalination proves to be the most economical method for increasing the Earth's freshwater supply, a demand that continues to surge daily.
- A solar still represents the most straightforward tool for acquiring freshwater from the plentiful reserves of salty water. Its functionality relies on the fundamental principles of solar distillation.
- Solar stills have limited productivity, and therefore, it is essential to increase the rate of heat transmission in order to increase the production rate.
- The two primary varieties of solar stills are active and passive stills. This categorization hinges on the heat source used to facilitate the evaporation process. While passive stills rely on internal heat sources within the still itself for powering the water evaporation process, active stills extract heat from external sources, such as industrial waste or solar collectors.
- The design of a solar still is uncomplicated, making it a cost-efficient tool for generating clean and drinkable water from brackish and saline water sources. This technology demonstrated significant advantages, particularly in arid and semiarid regions where the limited availability of drinking water is more pronounced compared to other places of the globe. Researchers are actively engaged in ongoing studies to enhance its efficiency and potentially scale up its production capabilities.

REFERENCES

Aadallah S, Abu-Khader MM, and Badarn OO. 2009. "Effect of various absorbing materials on the thermal performance of solar stills". *Desalination* 242(1–3):128–137.

Abdallah S, and Badran OO. 2003. "Sun tracking system for productivity enhancement of solar still". *Desalination* 44(9):1411–1418.

Abed FM, Ahmed AH, Hasanuzzaman M, Kumar L, and Hamaad NM. 2022. "Experimental investigation on the effect of using chemical dyes on the performance of single-slope passive solar still". *Solar Energy* 233:71–83.

Aboul-Enein S, El-Sebaii AA, and El-Bialy E. 1998. "Investigation of single-basin solar still with deep basins". *Renewable Energy* 14(1–4):299–305.

Abujazar MSS, Fatihah S, Ibrahim IA, Kabeel AE, and Sharil, S. 2018. "Productivity modelling of a developed inclined stepped solar still system based on actual performance and using a cascaded forward neural network model". *Journal of Cleaner Production* 170:147–159.

Adamu NI, Biswal DK, Pulagam MKR, and Rout SK. 2022. "Performance analysis of solar desalination system with thermal energy storage materials". *Materials Today: Proceedings* 74:801–807.

Al-Hinani H, Al-Nassri MS, and Jubran BA. 2002. "Effect of climate, design and operational parameters on the yield of a simple solar still". *Energy Conversion and Management* 43(13):1639–1650.

Ayber HS. 2006. "Mathematical modeling of an inclined solar water distillation system". *Desalination* 190(1–3):63–70.

Badran OO. 2007. "Experimental study of the enhancement parameters on single slope solar still productivity". *Desalination* 209:136–143.

Bagheri A, Esfandiari N, Honarvar B, and Azdarpour A. 2021. "ANN modeling and experimental study of the effect of various factors on solar desalination". *Journal of Water Supply: Research and Technology-Aqua* 70:41–57.

Bahiraei M, Nazari S, Moayedi H, and Safarzadeh H. 2020. "Using neural network optimized by imperialist competition method and genetic algorithm to predict water productivity of a nano fluid-based solar still equipped with thermoelectric modules". *Powder Technology* 366:571–586.

Bahiraei M, Nazari S, and Safarzadeh H. 2021. "Modeling of energy efficiency for a solar still fitted with thermoelectric modules by ANFIS and PSO enhanced neural network: A nano fluid application". *Powder Technology* 385:185–198.

Bassam AKA-H and Rababa'h HM. 2003. "Experimental study of a solar still with sponge cubes in basin". *Energy Conversion and Management* 44(9):677–688.

Benghanem M, Mellit A, Emad M, and Aljohani A. 2021. "Monitoring of solar still desalination system using the internet of things technique". *Energies*, 14:6892.

Bilal AA, Mousa SM, Omar O, and Uaser E. 1998. "Experimental evaluation of a single-basin solar still using different absorbing materials". *Renewable Energy* 14(1–4):307–310.

Biswal DK. 2021. "Performance assessment of a solar still using blackened surface". In Ramgopal, M., Rout, S.K., Sarangi, S.K. (eds) *Advances in Air Conditioning and Refrigeration*. Lecture Notes in Mechanical Engineering. Springer, Singapore.

Chen Z, Wei W, Ni BJ, and Chen H. 2022. "Plastic wastes derived carbon materials for green energy and sustainable environmental applications". *Environmental Functional Materials* 1:34–48.

Djebdedjian B, and Rayan MA. 2000. "Theoretical investigation on the performance prediction of solar still". *Desalination* 128(2):139–145.

Dwivedi VK, and Tiwari GN. 2006. "Annual Energy and energy analysis of single and double slope passive solar stills". *Trends in Applied Sciences Research* 3:225–241.

El-Bhai A, and Inan D. 1999. "Analysis of a parallel double glass solar still with separate condenser". *Renewable Energy* 17:509–521.

El-Sebaii AA, Aboul-Enein S, Ramadan MRI, and El-Bialy E. 2000. "Year-round performance of a modified single-basin solar still with mica plate as a suspended absorber". *Energy* 25(1):35–49

Essa, FA, Abd Elaziz M, and Elsheikh AH. 2020. "An enhanced productivity prediction model of active solar still using artificial neural network and Harris Hawks optimizer". *Applied Thermal Engineering* 170:115020.

Faegh M, Behnam P, Shafii MB, and Khiadani M. 2021. "Development of artificial neural networks for performance prediction of a heat pump assisted humidification-dehumidification desalination system". *Desalination* 508:115052.

Fath Hassan ES, and Hosny HM. 2002. "Thermal performance of a single-sloped basin still with an inherent built-in additional condenser". *Desalination* 142(1):19–27.

Feilizadeha M, Soltanieha M, Jafarpurb K, and Estahbanatic MRK. 2010. "A new radiation model for a single-slope solar still". *Desalination* 262(1–3):166–173.

Garg HP, and Mann HS. 1976. "Effect of climatic, operational and design parameters on the year round performance of single sloped and double sloped solar still under Indian arid zone conditions". *Solar Energy* 18:159–163.

Hanson A, Zachiritz W, Stevens K, Mimbela L, Polka R, and Cisneros L. 2004. "Discrete water quality of single basin solar still: Laboratory and field studies". *Solar Energy* 76(3):635–645

Jiang JY, Tian H, Cui MX, and Liu LJ. 2009. "Proof-of-concept study of an integrated solar desalination system". *Renewable Energy* 34(12):2798–2802

Kabeel AE. 2009. "Performance of solar stills with a concave wick evaporation surface". *Energy* 34(10):1504–1509.

Kabeel AE, El-Agouz, ES, Athikesavan MM, Duraisamy Ramalingam R, Sathyamurthy R, Prakash N, and Prasad C. 2020. "Comparative analysis on freshwater yield from conventional basin-type single slope solar still with cement-coated red bricks: An experimental approach". *Environmental Science and Pollution Research* 27:32218–32228.

Kandeal AW, An M, Chen X, Algazzar AM, Kumar Thakur A, and Guan X. 2021. "Productivity modeling enhancement of a solar desalination unit with nanofluids using machine learning algorithms integrated with bayesian optimization". *Energy Technology* 9:2100189.

Kaviti AK, Akkala SR, Ali MA, Anusha P, and Sikarwar VS. 2023a. "Performance improvement of solar desalination system based on CeO_2-MWCNT hybrid nanofluid". *Sustainability* 15:4268.

Kaviti AK, Akkala SR, Sikarwar VS, Snehith PS, and Mahesh M. 2023b. "Camphor-Soothed banana stem biowaste in the productivity and sustainability of solar-powered desalination". *Applied Science* 13:1652.

Khaled MS. 2010. "Improving the performance of solar still using vibratory harmonic effect". *Desalination* 251(1–3):3–11.

Khalifa AJN, Al-Jubouri AS, and Abed MK. 1999. "An experimental study on modified simple solar stills". *The Journal Energy Conversion and Management* 40(17):1835–1847.

Khanmohammadi S, and Khanjani S. 2021. "Experimental study to improve the performance of solar still desalination by hydrophobic condensation surface using cold plasma technology". *Sustainable Energy Technologies and Assessments* 45:101129.

Khanmohammadi S, Razi S, Delpisheh M, and Panchal H. 2023. "Thermodynamic modeling and multi-objective optimization of a solar-driven multi-generation system producing power and water". *Desalination* 545:116158.

Kumar A, Mary B, Siva A, and Kumari AA. 2021. "Influence of aluminium parabolic fins as energy absorption material in the solar distillation system". *Materials Today: Proceedings* 44:2521–2525.

Kumar VK, and Bai RK. 2008. "Performance study on solar still with enhanced condensation". *Desalination* 230(1–3):51–61.

Mahdi JT, Smith BE, and Sharif AO. 2011. "An experimental wick-type solar still system: Design and construction". *Desalination* 267:233–238.

Mashaly AF, and Alazba A. 2019. "Assessing the accuracy of ANN, ANFIS, and MR techniques in forecasting productivity of an inclined passive solar still in a hot, arid environment". *WSA* 45:239–250.

Mevada D, Panchal H, Ahmadein M, Zayed ME, Alsaleh NA, Djuansjah J, Moustafa EB, Elsheikh AH, and Sadasivuni KK. 2022. "Investigation and performance analysis of solar still with energy storage materials: An energy-exergy efficiency analysis. Case study". *Thermal Engineering* 29:101687.

Mirmanto I, Adi Sayoga M, Tri Wijayanta A, Pulung Sasmito A, and Muhammad AM. 2021. "Enhancement of continuous-feed low-cost solar distiller: Effects of various fin designs". *Energies* 14:4844.

Mohapatra A, Tejes PKS, Gembali C, and Kiran NB. 2022. "Design and performance analyses of evacuated U-tube solar collector using data-driven machine learning models". *ASME Journal of Solar Energy Engineering* 145(1):011007.

Moharana BR, and Sahoo SK. 2014. "An ANN and RSM integrated approach for predict the response in welding of dissimilar metal by pulsed Nd: YAG laser". *Universal Journal of Mechanical Engineering*, 2(5):169–173.

Moustafa SMA, and Brusewitz GH. 1979. "Direct use of solar energy for water desalination". *Solar Energy* 22(2):141–148.

Mowla D, and Karimi G. 1995. "Mathematical modelling of solar still in Iran". *Solar Energy* 55(5):389–393.

Murase K, Tobata H, Ishikawa M, and Tomaya S. 2006. "Experimental and numerical analysis of a tube-type networked solar still for desert technology". *Desalination* 190(1–3):137–146.

Murugavel KK, Sivakumar S, Ahmad JR, Chockalingam KK, and Srithar K. 2010. "Single basin double slope solar still with minimum basin depth and energy storing materials". *Applied Energy* 87(2):514–523.

Murugavel KK, and Srithar K. 2011. "Performance study on basin type double slope solar still with different wick materials and minimum mass of water". *Renewable Energy* 36(2):612–620.

Musharavati F, Khoshnevisan A, Alirahmi SM, Ahmadi P, and Khanmohammadi S. 2022. "Multi-objective optimization of a biomass gasification to generate electricity and desalinated water using Grey Wolf Optimizer and artificial neural network" *Chemosphere* 287(Part 2):131980.

Nafey AS, Mohamad MA, and Sharaf MA. 2008. "Enhancement of solar water distillation process by surfactant additives enhancement of solar water distillation process". *Desalination* 220(1–3):514–523.

Naim MM, and Mervat AE-K. 2003. "Non-conventional solar stills Part 2. Non-conventional solar stills with energy storage element". *Desalination*, 153(1–3):71–80.

Nebbia G, and Menozzi GN. 1966. "Historical aspects of dissalzione" *Acqua di Parma* 41:3–20.

Neel Shrimali VK Patel HP, and Sharma P. 2023. "Prediction of various parameters of desalination system using BOA-GPR machine learning technique for sustainable development: A case study". *Environmental Challenges* 12:100729.

Nijmeh S, Odeh S, Akash B. 2005. "Experimental and theoretical study of a single-basin solar still In Jordan". *International Communications in Heat and Mass Transfer* 32:565–572.

Omri A, Ofri J, and Nasrallah SB. 2005. "Natural convection effects in solar stills". *Desalination* 183(1–3):173–178.

Panchal H, Sadasivuni KK, Shanmugan S, and Pandya N. 2021. "Performance analysis of waste brick magnesia as a storage material in a solar still". *Heat Transfer* 50:1799–1811.

Prakash J, and Kavathekar AK. 1986. "Performance prediction of a regenerative solar still". *International Journal of Solar & Wind Technology* 3(2):119–128.

Saba S, Bahrami A, Delfani F, and Sohani A. 2023. "Uncertainty covered techno-enviro-economic viability evaluation of a solar still water desalination unit using Monte Carlo approach". *Energies* 16(19):6924.

Sahoo BB, Sahoo N, Mahanta P, Borbora L, Kalitha P, and Saha UK. 2008. "Performance assessment of solar still using black ended surface and thermocol insulation". *Renewable Energy* 33(1):1703–1708.

Salem H, Kabeel AE, El-Said EMS, and Elzeki OM. 2022. "Predictive modelling for solar power-driven hybrid desalination system using artificial neural network regression with Adam optimization" *Desalination* 522:115411.

Shakthivel M, and Shanmugasundaram S. 2008. "Effect of energy storage medium on the performance of solar still". *International Journal of Energy Research* 32(1):68–82.

Shukla S, and Sorayan VPS. 2005. "Thermal modelling of solar stills: An experimental validation". *Renewable Energy* 30(5):683–690.

Shukla SK. 2003. "Computer modelling of passive solar still by evaluating absorptivity of basin liner". *International Journal of Ambient Energy* 24(3):123–132.

Sohani, A, Hoseinzadeh, S, Samiezadeh, S, and Verhaert I. 2022. "Machine learning prediction approach for dynamic performance modeling of an enhanced solar still desalination system". *Journal of Thermal Analysis and Calorimetry* 147: 3919–3930.

Tanaka H, and Nakatake Y. 2006. "Theoretical analysis of a basin type solar still with internal and external reflectors". *Desalination* 197(1–3):205–216.

Tiwari AK, and Tiwari GN. 2006. "Effect of water depths on heat and mass transfer in a passive solar still: In summer climatic condition". *Desalination* 195(1–3):78–94.

Torchia-Núñeza JC, Porta-Gándarab MA, and Cervantes-de Gortaria JG. 2008. "Energy analysis of a passive solar still". *Renewable Energy* 33(4):608–616.

Valsaraj P. 2002. "An experimental study on solar distillation in a single slope basin still by surface heating the water mass". *Renewable Energy* 25(4):607–612.
Velmurugan V, Gopalakrishnan M, Raghu R, and Srithar K. 2008. "Single basin solar still with fin for enhancing productivity". *Energy Conversion and Management* 49(10):2602–2608.
Vigneswaran VS, Suresh Kumar P, Kumar PG, Aravind Kumar J, Siva Chandran S, Kumaresan G, and Shanmugam M. 2023. "Enhancement of passive solar still yield through impregnating water jackets on side walls – A comprehensive study" *Solar Energy* 262:111841.
Voropoulos K, Mathioulakis E, and Belessiontis V. 2002. "Analytical simulation of energy behaviour of solar stills and experimental validation". *Desalination* 153(2):87–94.
Wang Y, Kandeal AW, Swidan A, Sharshir SW, Abdelaziz GB, and Halim MA. 2021. "Prediction of tubular solar still performance by machine learning integrated with bayesian optimization algorithm". *Applied Thermal Engineering* 184:116233.
Yadav YP, and Kumar A. 1991. "Transient analytical investigations on a single basin solar still with water flow in the basin". *Energy Conversion Management* 31(1):27–38.
Yeh H. 1993. "Experimental studies on upward-type double effect solar distillers with air flow through the second effect". *Energy* 18(11):1107–1111.
Zarei T, and Behyad R. 2019. "Predicting the water production of a solar seawater greenhouse desalination unit using multi-layer perceptron model". *Solar Energy* 177:595–603.
Zeinab SAR, and Ashraf L. 2007. "Experimental and theoretical study of a solar desalination system located in Cairo, Egypt". *Desalination* 217(1–3):52–64.
Zouli N. 2023. "Design of solar power-based hybrid desalination predictive method using optimized neural network" *Desalination* 566:116854.

3 Statistical Modelling for Ergonomic Assessment of Lathe Machine Operators

A Sustainable Approach to Assess the Impact of Occupational Health Hazards

Shailesh Dewangan and Chitrakant Tiger
Chouksey Engineering College, Bilaspur, India

B. S. Chawla
Government Engineering College, Bilaspur, India

Bikash Ranjan Moharana
Papua New Guinea University of Technology, Lae, Papua New Guinea

3.1 INTRODUCTION

The knowledge of health, hygiene, safety, and ergonomics applications are essential for industrial and societal advancement. Happiness in the workplace may be elusive without an understanding of labour welfare services. Workplace intervention is not possible without taking OHS/ergonomics into account. Musculoskeletal disorders (MSDs) during work are recognized as a primary, prevalent, and severe occupational health concern (Nambiema et al., 2020). MSDs are defined as injuries or abnormalities of the muscles, nerves, tendons, joints, cartilage, and spinal discs by the Centers for Disease Control and Prevention. On the other hand, work-related musculoskeletal disorders (WMSDs) are conditions where the tasks and work environment

DOI: 10.1201/9781003349877-3

considerably exacerbate the condition, particularly if it endures for a lengthy period of time (Manuele, 2007; Walder et al., 2007; Saurin et al., 2008). Work conditions that may cause WMSDs include repetitive large object lifting, body vibration, monotonous overhead work, working with the neck in chronic flexion, and performing repetitive forceful tasks. The level of exposure to risk factors for WMSDs can be used to develop and carry out an interventional ergonomics program in the workplace.

A wide variety of surveys and analysis were conducted on MSDs by various organizations across the globe to find a better and healthier work environment. MSD accounts for 31% of all work-related injuries and illnesses in the United States, according to studies conducted by the U.S. Bureau of Labour Statistics in 2015, which includes overexertion of the body (U.S. Bureau of Labor Statistics, 2016). According to an analysis conducted in 2018 by BC Legal on Workplace Health and Safety, MSDs comprised almost 35% of work-related health issues in the United Kingdom. Australian Bureau of Statistics (2018) research shows that trade workers and technicians are majorly affected by MSDs, nearly 24%. Godrej & Boyce, the flagship company of the Godrej Group, reported that almost 76% of the workforce is affected by WMSDs, which includes the employees of MNCs and Indian corporates. They are mostly office-going workers (BioSpectorm, 2022). According to a 2019 report by Brazil's Special Secretariat for Social Security and Labour, nearly 39,000 workers were removed from their jobs each year as a result of WMSDs. This results in a loss of functionality due to the worker's difficulty moving around, which affects their personal and professional lives (Fundacentro, 2020). These MSDs are not only affecting the worker's health but also have an economic impact on businesses, industries, and the nation as a whole, as billions is spent annually on direct compensation for MSDs (Muduli et al., 2023). Workplace health and safety will not occur without a conscious understanding of work regulation and labour law; nevertheless, implementation of work regulation and labour legislation will be difficult without the sincere commitment of the local government authority or regulatory agency. Transparency of regulatory agencies is impossible to achieve without the establishment of a neutral watchdog group, such as the National Health and Safety Council (Manuele, 2007).

Musculoskeletal challenges may differ on the sort of activity or job description assigned to an individual. Standing and sitting are two of the most popular postures during any work. Standing posture is typically preferred in many industries due to leg mobility and a high degree of freedom. These working postures play an important role for physical and mental stability. Mechanical (physical) exposure at work may generate WMSDs, but psychosocial factors at work may create pain or affect the perceived pain level caused by mechanical exposure (Westgaard and Winkel, 2011).

Lathe machine operation is one the most common machining processes in many manufacturing industries. Different hazards like flying debris, dust, cutting/crushing due to sharp tool, noise, vibration, and so on, could be minimized through several measures such as personal protective equipment, machine guards, timely inspection and maintenance of the tool, dust collector/proper extraction systems, adequate training, and so forth. This information is provided in the machining areas for worker awareness, but offers very limited awareness regarding MSDs. It has been observed that the ligaments of the shoulder and low back muscles are the most commonly affected muscles among lathe machine workers, as they deal mainly with prolonged spinal cord bending, inadequate lower back support, repetitive lifting, differential lifting height, ambient circumstances, long working hours, and so on. MSDs are conditions in which a portion of the musculoskeletal system becomes injured over time as a result of repetitive tasks (Sachdeva et al., 2011). Musculoskeletal diseases are conditions in which the human body is traumatized in a minor or major manner over time. Usually, workers tend to overexert their bodies to such an extent that it becomes difficult for them to withstand the soreness they feel after some time (Andersen et al., 2007).

Researchers are using a number of ergonomics analysis approaches to evaluate the threats associated with WMSD. Using a wide-ranging questionnaire survey in manufacturing companies, Probst (2004) has claimed that safety performance is comprised of elements including climate safety, support from management, risk management, proper safety communication, competency in employees' safety, and safety training. According to DeJoy et al. (2004), staffs' attitude is a major concern in safety issues. They have also emphasized that industrial mishaps have an impact on human capital in addition to causing financial losses because they can interrupt industrial processes, damage manufacturing equipment, and tarnish the company's brand. Although the physical risk variables discussed in Huang et al.'s (2003) research are significant first-line risk factors, other plausible elements, such as organizational and psychological factors, may either directly or indirectly cause a condition or affect the impact of physical risk factors. Three categories of risk variables have been identified: biomechanical exposures, psychosocial stressors, and individual risk factors. Poorly built workplaces and biomechanical exposures to strong forces, repeated action, and deviations from neutral body alignments are examples of biomechanical exposures. Khan et al. (2020) provides an overview of the fundamentals of ergonomic evaluation utilizing the University of Michigan's Energy Expenditure Prediction Programme (EEPP) software to maximize worker productivity, efficiency, and safety in the context of manufacturing. Using software called the EEPP, Mohsin et al. (2021) looked into energy expenditure, that is, the mean metabolic energy rate was estimated. By monitoring and documenting the actions taken during a lathe machine's operations, the energy consumption was ascertained.

Liu et al. (2019) proposed an ergonomic design principle for the CNC lathe machine tool to create a good man-machine interaction and effectively improve the operator's work, so better ensuring the ease, comfort, and safety of the operator. Starting with the CNC lathe's appearance design, the paper improves the design of a few functional parts for the original equipment using the ergonomic design concept and approach. Hernandez Arellano et al. (2017) investigated relationships between musculoskeletal discomfort, fatigue, and workload, as well as differences in workload and fatigue items/dimensions connected to job and demographic characteristics for the CNC lathe machine tool. They also suggested that the workers' sleeping hours, physique, weight handled, and so on, should also considered for MSDs analysis, which may aid in the development of effective future interventions. Seppälä and Tuominen (1992) provided a comparison statement on physical and mental strain encountered by conventional and advanced manufacturing technology (i.e., CNC and

FMS) operators. Environmental working conditions in advanced systems were shown to be more convenient than traditional ones. Researchers are using several ergonomics analysis concepts, tools, and features to evaluate the challenges associated with WMSDs. Based upon the literatures, the study is being conducted with the following objectives:

- To determine MSD risk factors during lathe operation using ergonomics analysis tools.
- To suggest factors for reducing the risk of MSDs in the machining workshop.

The following research questions (RQs) represent the framework for the present study:

RQ 1. What is the correlation between WMSD symptoms and lathe operators?
RQ 2. What relationship exists between the age of the lathe operators and the symptoms of WMSDs?
RQ 3. How is the MCDM approach helpful for determining the best ergonomics posture?

By learning how to control risk variables, employers can lower occupational injuries while simultaneously increasing worker comfort, productivity, and job satisfaction. Workers should be informed by their employers about easy workplace modifications that can be done to delay the onset of MSDs. WMSDs can be avoided by putting an efficient ergonomics plan into practice, which leads to the sustainable development of mankind.

3.2 EXPERIMENTAL DESIGN

3.2.1 Selection of Process Parameters

A wide range of illnesses that involve the bones, joints, muscles, and connective tissues are referred to as musculoskeletal disorders. These illnesses ranked among the most expensive and incapacitating conditions worldwide and may lead to pain and loss of functionality. The prevention of WMSDs is now highly essential in industries, which ultimately leads to the workforce and economic growth of the nation. A limited but wide variety of analysis was conducted on WMSDs by different researchers worldwide to find a better and healthier work environment in industries and workplaces. Sachdeva et al. (2011) made an approach to minimize the MSDs for workers working in the manufacturing units using fuzzy logic. Nourmohammadi et al. (2023) suggested a multi-criteria decision making (MCDM) approach, that is, an enhanced non-dominated sorting genetic algorithm (E-NSGA-II) may ensure the selection of promising solutions for worker posture, which are in line with ergonomic factors (Industry 5.0). Kamath et al. (2020) investigated the prevalence of MSDs among the operators who work in the mechanical engineering laboratories via evolutionary algorithms such as rapid entire body assessment and rapid upper limb assessment.

In the present study, the average metabolic energy rate is predicted by knowing the energy expenditure and task duration. Increasing productivity and the quality of the product are the main challenges of manufacturing industries. L_{27} experiment-based orthogonal array design was used to study operator ergonomics in machining. Based on L_{27}, orthogonal array design performs the experiment by operator and notes responses from the operator. Output responses are optimized by a MCDM approach, that is, the grey relation method, which gives minimum pain in posture at machining. Significant effects of all responses are found to show that these parameters are more critical in posture at machining.

This research is conducted at a machining workshop with different temperature conditions. Conveniently for study, different operators were selected, of average height 165–175 cm, with three different age factors: 22 years, 35 years, and 48 years. During the experiment, workers' movements were recorded, along with their responses to the pain in their legs, hands, and backs, marked at a range between 0 and 9. The three input parameters are taken for design the experiment are age factor, working time, standing position and working temperature. In this experiment similar types of machine were used to perform the experiment and having three levels of each factor was taken as input parameter as shown in Table 3.1.

For design the performance standing position are converting in numeric number are shown as follow:

- Straight: 1
- Slightly bent: 2
- Down: 3

Those factor that affect the operator are shown in Figure 3.1: age, continuous working time, working temperature, standing position, and so on.

3.2.2 Method Flow Process

The conceptual framework of ergonomics optimization is shown in Figure 3.2. It provides an outline of potential safety causes, offers relief from severe pain, and illustrates a recommended method for an occupational safety and health management system to protect employees from risks related to their jobs.

Design matrix is a powerful statistical analysis pathway, in which the influence of the elements with the product's design specifications could be estimated. Through

TABLE 3.1
Factors and Their Levels

Parameters	Symbol	Units	Levels		
			L1	L2	L3
Age	A	Years	22	35	48
Working time	W_t	Hr	3	3.5	4
Standing position	S_p	---	Straight	Slightly bend	Down
Working temperature	T	°C	28	32	36

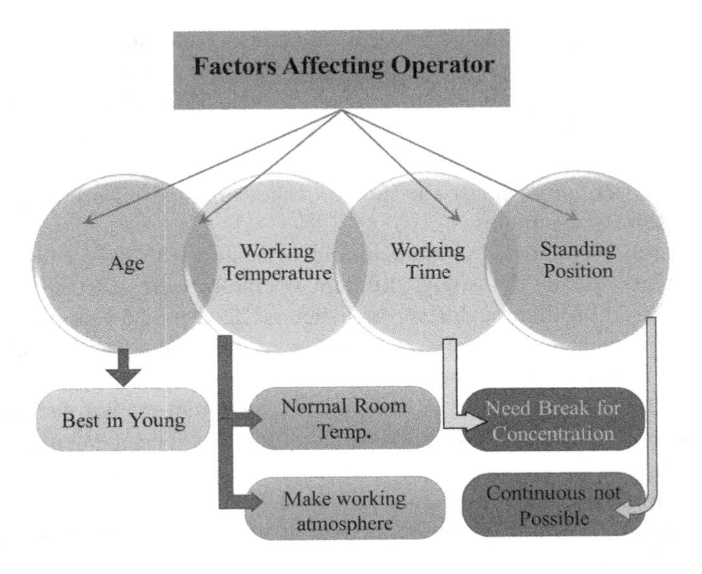

FIGURE 3.1 Factors affecting operator.

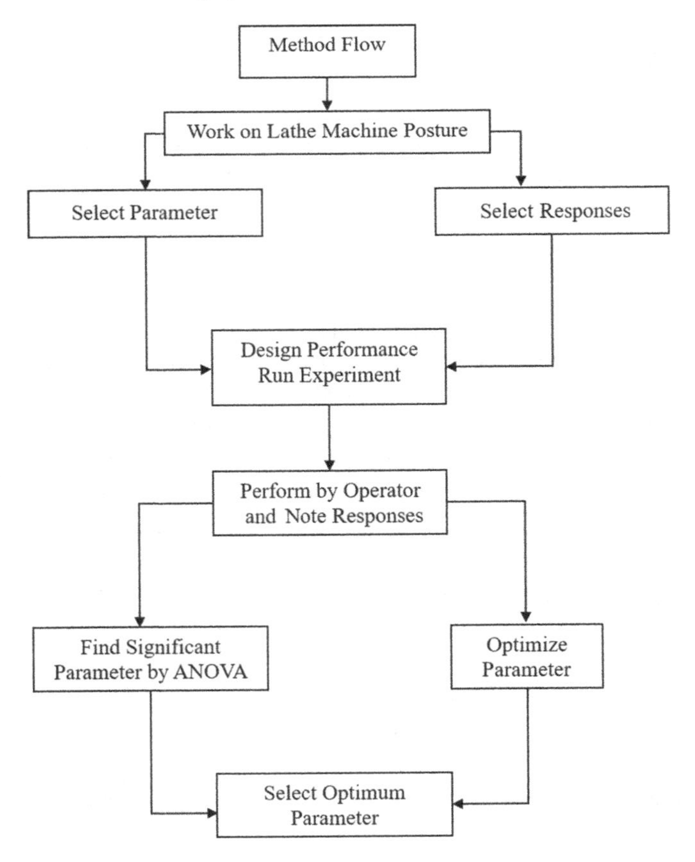

FIGURE 3.2 Method flow chart.

this analysis, management can efficiently trade off the less significant qualities for the more significant ones by knowing these linkages. Several researchers are following different statistical models in order to address complex challenges effectively. In comparison to other designs, it has been noted that the Taguchi orthogonal array is one of the most effective and efficient design matrices (Ross, 1996; Biswal et al., 2022; Moharana et al., 2023), which is successfully implemented in various manufacturing processes. With orthogonal arrays, a standard experimental design, the significant impact of an element on the result can be determined with a minimal number of experimental trials. Hence, the Taguchi orthogonal design matrix was considered for the current ergonomic analysis. The analysis considered four input process parameters (age, working time, standing position, and working temperature) with three different levels each (as given in Table 3.1), followed by 27 experimental trials (L_{27}) as presented in Table 3.2. Based on the main factor, the variables are assigned to columns, as stipulated by orthogonal array.

TABLE 3.2
Performance Parameter and Responses by Operators

Sl. No.	Input Parameters				Responses by Operators		
	A	W_t	S_p	T	Leg Pain (L)	Hand Pain (H)	Back Pain (B)
1	22	3	1	28	3	4	2
2	22	3	2	28	1	3	4
3	22	3	3	28	2	4	6
4	22	3.5	1	32	4	5	2
5	22	3.5	2	32	2	5	3
6	22	3.5	3	32	3	5	5
7	22	4	1	36	6	8	4
8	22	4	2	36	4	7	6
9	22	4	3	36	5	9	8
10	35	3	1	36	6	9	4
11	35	3	2	36	6	7	5
12	35	3	3	36	8	6	8
13	35	3.5	1	28	5	6	6
14	35	3.5	2	28	3	2	5
15	35	3.5	3	28	6	5	6
16	35	4	1	32	4	7	3
17	35	4	2	32	3	6	5
18	35	4	3	32	5	7	4
19	48	3	1	32	6	8	5
20	48	3	2	32	5	5	6
21	48	3	3	32	7	8	8
22	48	3.5	1	36	8	8	5
23	48	3.5	2	36	7	9	6
24	48	3.5	3	36	8	3	8
25	48	4	1	28	7	5	4
26	48	4	2	28	6	4	8
27	48	4	3	28	5	6	6

3.2.3 Consideration of Output Responses

Responses are obtained on the basis of the performance run by the operator. After completing the performance, the operator gives the report of their leg, hand, and back pain in terms of a number from zero to nine. All optimization was done on the basis of operator marking. Zero marks suggests no pain is there, whereas nine marks is considered heavy pain. Each performance is run by the operator after rest position; the operator will not perform their experiment continuously. Some of the positions obtained by operators are shown in Figure 3.3–3.5 with respect to age and standing position.

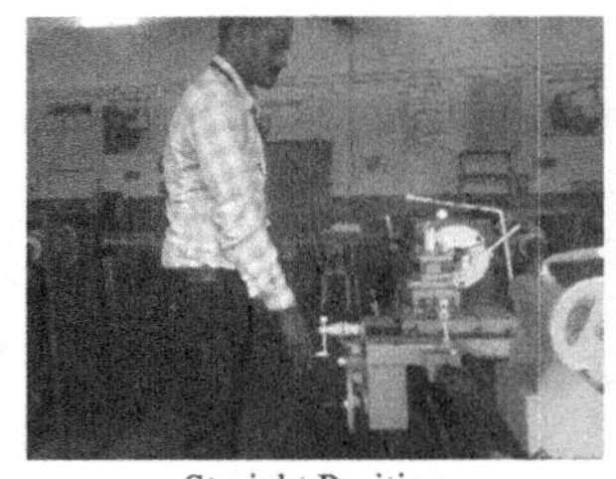

Straight Position

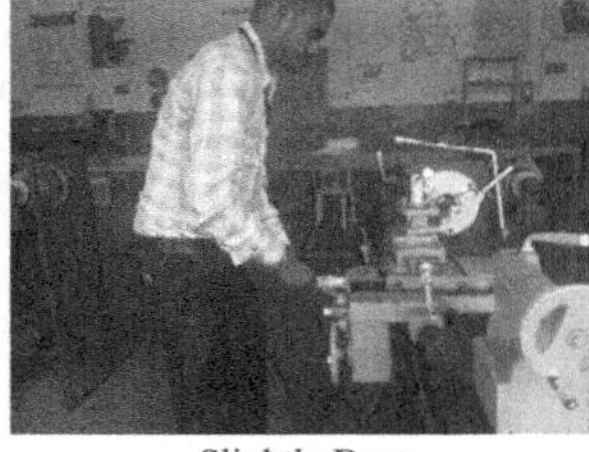

Slightly Bent

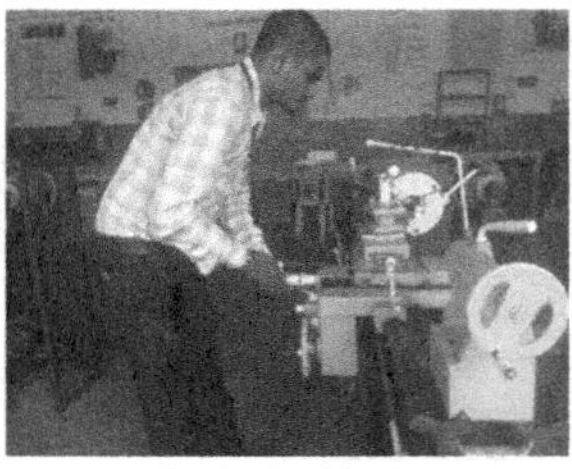

Down Position

FIGURE 3.3 Operator 22 years old in three positions.

Straight Position

Slightly Bent

Down Position

FIGURE 3.4 Operator 36 years old in three positions.

Straight Position

Slightly Bent

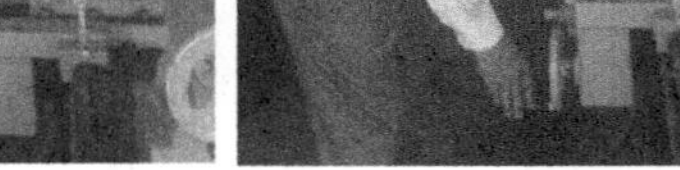

Down Position

FIGURE 3.5 Operator 48 years old in three positions.

3.3 RESULTS & DISCUSSIONS

3.3.1 Result Analysis

As per the design, performances were obtained showing that an increase in working temperature in fact increases the stress of operator. The pain level experienced by the operators expressed in weights as a decision maker is given in Table 3.3. On the base of design of experiment (DOE) parameter, the performance chart is followed by each operator. Their responses are shown in Table 3.2.

3.3.2 Optimization of Process Parameters

In the present study, three WMSDs output responses are recorded from the operators during lathe operations: leg pain, hand pain, and back pain. In order to optimize the current situation, an MCDM approach, that is, grey relational analysis (GRA) was be implemented. GRA is a mathematical model that converts several responds to a single response based on grey system theory and has been also effectively analyzed for different research fields by several researchers (Kirubakaran and Ilangkumaran, 2016; Tsaur et al., 2017; Moharana and Patro, 2019). For ease of optimization, multiple outcomes are combined into a single response in this type of analysis, known as the grey relational grade (GRG). The output responses were first normalized based on the response characteristics (as indicated in Table 3.4), followed by the grey relational coefficients for each process response, as shown in Table 3.4. The GRG was then calculated by averaging the grey relational coefficient values. The GRG serves as the foundation for the overall assessment of the multiple responses, according to the GRG rules "Higher is better" policy. The greatest value of GRG has optimum parameters for experimented responses as shown in Table 3.4.

Analysis of Variance (ANOVA) is a tool that determines the impact of process variables associated with the response. Furthermore, the table quantitatively describes the impact of each input process parameter on the overall quality attributes of the process, demonstrating the applicability of the numerical model. The literature review indicates that the design should have a 95% confidence level. In other words,

TABLE 3.3
Decision Maker

Pain Categories	Respected Weighted
Very tiny Pain	1
Tiny pain	2
Very small pain	3
Small pain	4
Medium pain	5
Large pain	6
Very large pain	7
Huge pain	8
Very huge pain	9

TABLE 3.4
Normalization of Responses, Grey Relation Coefficient and GRG Values

Sl. No.	Normalization of Responses (X_i)			Grey Relation Coefficient (G_i)			GRG (Γ)
	(L)	(H)	(B)	(L)	(H)	(B)	
1	0.71429	0.71429	1.00000	0.63636	0.63636	1.0000	0.75758
2	1.00000	0.85714	0.66667	1.00000	0.77778	0.6	0.79259
3	0.85714	0.71429	0.33333	0.77778	0.63636	0.42857	0.61424
4	0.57143	0.57143	1.00000	0.53846	0.53846	1.00000	0.69231
5	0.85714	0.57143	0.83333	0.77778	0.53846	0.75	0.68875
6	0.71429	0.57143	0.5	0.63636	0.53846	0.5	0.55828
7	0.28571	0.14286	0.66667	0.41177	0.36842	0.6	0.46006
8	0.57143	0.28571	0.33333	0.53846	0.41177	0.42857	0.4596
9	0.42857	0	0	0.46667	0.33333	0.33333	0.37778
10	0.28571	0	0.66667	0.41177	0.33333	0.6	0.44837
11	0.28571	0.28571	0.5	0.41177	0.41177	0.5	0.44118
12	0	0.42857	0	0.33333	0.46667	0.33333	0.37778
13	0.42857	0.42857	0.33333	0.46667	0.46667	0.42857	0.45397
14	0.71429	1.00000	0.5	0.63636	1.00000	0.5	0.71212
15	0.28571	0.57143	0.33333	0.41177	0.53846	0.42857	0.4596
16	0.57143	0.28571	0.83333	0.53846	0.41177	0.75	0.56674
17	0.71429	0.42857	0.5	0.63636	0.46667	0.5	0.53434
18	0.42857	0.28571	0.66667	0.46667	0.41177	0.6	0.49281
19	0.28571	0.14286	0.5	0.41177	0.36842	0.5	0.42673
20	0.42857	0.57143	0.33333	0.46667	0.53846	0.42857	0.4779
21	0.14286	0.14286	0	0.36842	0.36842	0.33333	0.35673
22	0	0.14286	0.5	0.33333	0.36842	0.5	0.40059
23	0.14286	0	0.33333	0.36842	0.33333	0.42857	0.37678
24	0	0.85714	0	0.33333	0.77778	0.33333	0.48148
25	0.14286	0.57143	0.66667	0.36842	0.53846	0.6	0.50229
26	0.28571	0.71429	0	0.41177	0.63636	0.33333	0.46049
27	0.42857	0.42857	0.33333	0.46667	0.46667	0.42857	0.45397

TABLE 3.5
Analysis of Variance for GRG

Source	DF	Seq SS	F	P	% Contribution	Rank
A	1	0.138969	47.9578	0.000001	43	1
W_t	1	0.014020	4.8384	0.038636	4	4
S_p	1	0.069629	24.0288	0.000067	22	3
T	1	0.098437	33.9702	0.000007	31	2
Error	22	0.063750				
Total	26				100	

S = 0.0538306 R^2 = 83.3% R^2 (adj) = 80.42%

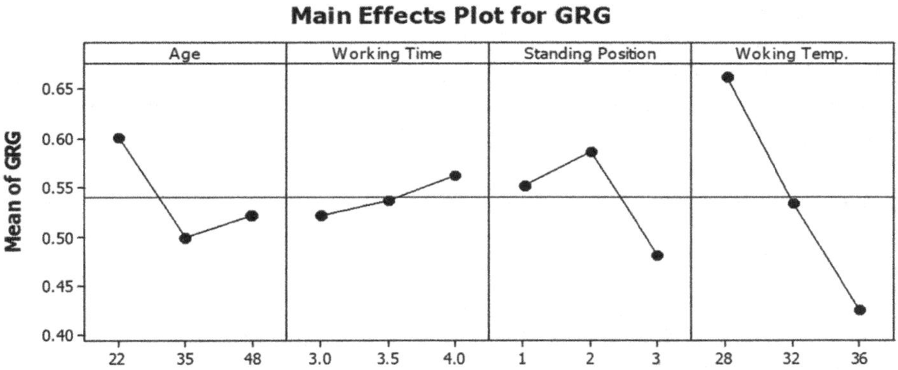

FIGURE 3.6 Main effect plot for GRG.

if the probability (P) value is lower than 0.05, the factor is considered significant. The F value can be used to evaluate which factor has a major impact on the performance characteristic. ANOVA table for GRG is shown in Table 3.5. It is evident from the table that each input variable has a substantial impact on the responses. But age is the most important factor, followed by temperature, position, and working time. Table 3.5 also displays the proportion of contribution and the corresponding rank of the variables. The R^2 value indicates the greatest adequacy of the model with respect to output response. The main effect plot (Figure 3.6) shows the parameter that has higher responses gives the better result for combined optimum pain.

3.4 IMPLICATION AND FUTURE RECOMMENDATION

Prevention of injuries and MSDs during work is the aim of ergonomics. It's essential to recognize the physical strain that non-ergonomic work positions impose on the human body. The current method for ergonomic analysis during lathe operation is based upon the case studies and statistical analysis simultaneously in order to optimize human well-being and overall system performance. This technique may provide an awareness for laboratory demonstrators and machine operators in educational institutions. The digital technology implementation includes a cloud-based platform, sensors, image processing, and so on, so that ergonomics practices may provide a quick view into all assessments, improvements, and direct causes at our fingertips. Manual observation was traditionally used to conduct ergonomics assessments. This method has shortcomings such as errors and discrepancies among assessments and it is time consuming. Artificial intelligence is used in computer vision technology, and deep learning allows it to see what the human eye cannot in fractions of a second.

3.5 CONCLUSIONS

The MCDM approach was used to analyse body posture for lathe machine operators in order to identify the major posture of individuals who are working in uncomfortable and unpleasant positions. Based on the observation, the following conclusions can be drawn on the effect of operators performing machining operations who gave their individual responses in terms of leg pain, hand pain, and back pain. The main effect plot of GRG ergonomics posture parameter for lathe operators proposed the least pain condition is age 22 years, working time is 4 hours, standing position is slightly bent, and working temperature is 28°C.

REFERENCES

Andersen, J.H., Haahr, J.P. and Frost, P., 2007. Risk factors for more severe regional musculoskeletal symptoms: A two-year prospective study of a general working population. *Arthritis & Rheumatism*, *56*(4), pp. 1355–1364.

Australian Bureau of Statistics (2018) *Work Related Injuries*, Retrieved from Australian Bureau of Statistics

BC Legal (2018) *Health and Safety at Work Statistics 2017/18 Published*, 8 November, from BC Legal: https://www.bc-legal.co.uk/bcdn/733-254-health-and-safety-at-work-statistics-2017-18-published.html

BioSpectorm, 76% workforce suffers from work-related musculoskeletal disorders: Godrej study, Published on 28/10/2022. https://www.biospectrumindia.com/news/98/22182/76-workforce-suffers-from-work-related-musculoskeletal-disorders-godrej-study.html

Biswal, D.K., Moharana, B.R. and Mohapatra, T.P., 2022. Bending response optimization of an ionic polymer-metal composite actuator using orthogonal array method. *Materials Today: Proceedings*, *49*, pp. 1550–1555.

DeJoy, D.M., Gershon, R.R. and Schaffer, B.S., 2004. Safety climate: Assessing management and organizational influences on safety. *Professional Safety*, *49*(7), p. 50.

Fundacentro and SRTs held debates on how to prevent these illnesses, Published on 03/11/2020 00:00 Updated on 08/17/2022 17:59 By Social Communication Service - SCS / Fundacentro. https://www.gov.br/fundacentro/pt-br/assuntos/noticias/noticias/2020/3/

Hernandez Arellano, J.L., Serratos Perez, J.N., Alcaraz, J.L.G. and Maldonado Macias, A.A., 2017. Assessment of workload, fatigue, and musculoskeletal discomfort among computerized numerical control lathe operators in Mexico. *IISE Transactions on Occupational Ergonomics and Human Factors*, *5*(2), pp. 65–81. https://www.abs.gov.au/statistics/labour/earnings-and-work-hours/workrelated-injuries/latest-release#data-download (accessed 30th October 2018).

Huang, G.D., Feuerstein, M., Kop, W.J., Schor, K. and Arroyo, F., 2003. Individual and combined impacts of biomechanical and work organization factors in work-related musculoskeletal symptoms. *American Journal of Industrial Medicine*, *43*(5), pp. 495–506.

Kamath, C.R., Naik, N., Bhat, R., Mulimani, P. and Sinniah, A., 2020. Assessing the possibility of musculoskeletal disorders occurrence in the mechanical engineering laboratory operators of educational institutes. *International Journal of Advanced Science and Technology*, *29*(3), pp. 6191–6197.

Khan, Y., Khan, S. and Naushad, O., 2020. Ergonomics evaluation for the operation performed on Lathe machine using energy expenditure prediction program. *International Research Journal of Engineering and Technology*, 7(8), pp. 4181–4186.

Kirubakaran, B. and Ilangkumaran, M., 2016. Selection of optimum maintenance strategy based on FAHP integrated with GRA–TOPSIS. *Annals of Operations Research*, *245*, pp. 285–313.

Liu, L., Li, W., Tan, Z., Zhang, M. and Bo, M., 2019. Model design and analysis improvement of CNC Lathe Based on Ergonomics. In *IOP Conference Series: Materials Science and Engineering* (Vol. 573, No. 1, p. 012045). IOP Publishing.

Manuele, F.A., 2007. Lean concepts opportunities for safety professionals. *Advanced Safety Management Focusing on Z10 and Serious Injury Prevention*, *Journal of the American Society Safety Professionals*, *52*(08), pp. 255–269.

Moharana, B.R. and Patro, S.S., 2019. Multi objective optimization of machining parameters of EN-8 carbon steel in EDM process using GRA method. *International Journal of Modern Manufacturing Technologies*, *11*(2), pp. 50–56.

Moharana, B.R., Mohapatra, K.D., Muduli, K., Biswal, D.K. and Moharana, T.K., 2023. Multi-response optimisation of machining parameters in WEDM using hybrid desirability-based TOPSIS concept. *International Journal of Process Management and Benchmarking*, *14*(4), pp. 439–459.

Mohsin, S., Khan, I.A. and Ali, M., 2021. Role of Energy Expenditure in the Evaluation of Lathe Operator Performance. *International Journal for Modern Trends in Science and Technology*, 7, 0708068, pp. 203–212.

Muduli, K., Syed, S.A., Kommula, V.P., Moharana, B.R. and Behera, B.C., 2023. Assessment of musculoskeletal disorder in food service industry in emerging economies. *International Journal of Process Management and Benchmarking*, *14*(4), pp. 558–580.

Nambiema, A., Bodin, J., Fouquet, N., Bertrais, S., Stock, S., Aublet-Cuvelier, A., Descatha, A., Evanoff, B. and Roquelaure, Y., 2020. Upper-extremity musculoskeletal disorders: How many cases can be prevented? Estimates from the COSALI cohort. *Scandinavian Journal of Work, Environment & Health*, *46*(6), p. 618.

Nourmohammadi, A., Ng, A.H., Fathi, M., Vollebregt, J. and Hanson, L., 2023. Multi-objective optimization of mixed-model assembly lines incorporating musculoskeletal risks assessment using digital human modeling. *CIRP Journal of Manufacturing Science and Technology*, *47*, pp. 71–85.

Probst, T.M., 2004. Safety and insecurity: exploring the moderating effect of organizational safety climate. *Journal of Occupational Health Psychology*, *9*(1), p. 3.

Ross, P.J., 1996. *Taguchi Techniques for Quality Engineering*, McGraw-Hill Education, Australia.

Sachdeva, A., Gupta, B.D. and Anand, S., 2011. Minimizing musculoskeletal disorders in lathe machine workers. *International Journal of Ergonomics*, *1*(2), pp. 20–28.

Saurin, T.A., Formoso, C.T. and Cambraia, F.B., 2008. An analysis of construction safety best practices from a cognitive systems engineering perspective. *Safety Science*, *46*(8), pp. 1169–1183.

Seppälä, P. and Tuominen, E., 1992. Job characteristics and mental and physical well-being experienced by the operators of CNC, FMS and conventional machine tools. *International Journal of Industrial Ergonomics*, *9*(1), pp. 25–35.

Tsaur, R.C., Chen, I.F. and Chan, Y.S., 2017. TFT-LCD industry performance analysis and evaluation using GRA and DEA models. *International Journal of Production Research*, *55*(15), pp. 4378–4391.

U.S. Bureau of Labor Statistics (2016) *Nonfatal Occupational Injuries and Illnesses Requiring Days Away from Work, 2015*, 10 November, U.S. Bureau of Labor Statistics [online] https://www.bls.gov/news.release/osh2.nr0.htm

Walder, J., Karlin, J. and Kerk, C., 2007. Integrated Lean Thinking & Ergonomics: Utilizing Material Handling Assist Device. *Solutions for a productive workplace. An MHIA White paper. Material Handling Industry of America Nov.*

Westgaard, R.H. and Winkel, J., 2011. Occupational musculoskeletal and mental health: Significance of rationalization and opportunities to create sustainable production systems–A systematic review. *Applied Ergonomics*, *42*(2), pp. 261–296.

4 An Evaluation of the Impact of Circular Economy (CE) Models Based on AI and IoT for Job Creation and Reallocation

Arun Kumar Singh and Benson Mirou
Papua New Guinea University of Technology, Lae, Papua New Guinea

4.1 INTRODUCTION

As circular economy (CE) models progress with artificial intelligence (AI) and Internet of Things (IoT), the workforce may require additional skills to fully harness the potential of these technologies. This transition presents possibilities for labor reallocation as individuals adapt to evolving roles. The collaboration between CE, AI, and IoT can stimulate economic growth by reducing expenses, fostering innovation, and attracting investments in sustainable technologies. These economic benefits can translate into employment creation across various sectors. CE models driven by AI and IoT offer a fertile environment for entrepreneurial endeavors. Startups that focus on sustainable products, materials, and technologies contribute to employment generation and innovation. Governments have a vital role in influencing the impact of CE models. Supportive policies can incentivize businesses to adopt sustainable practices, potentially leading to job growth in environmentally friendly industries. The global nature of CE models means that their influence extends beyond national borders. International collaboration and competition can impact job markets on a global scale. This assessment recognizes that while CE models based on AI and IoT have the potential to generate jobs and redistribute labor, there are challenges to address, including bridging the digital divide and ensuring responsible usage of AI to avoid unintended consequences. It highlights the significance of a comprehensive approach, where governments, industries, and educational institutions collaborate to maximize the positive effects of these technologies on employment creation and redistribution within the framework of a circular economy.

DOI: 10.1201/9781003349877-4

Research Questions:

- What techniques may be used to achieve a positive and equitable outcome for the workforce in the face of these technological developments and how do emerging technologies, particularly AI and automation, effect job creation and reallocation in diverse industries?
- What effects on job creation and reallocation does the integration of AI and the IoT into CE models have across different industries and economies?
- How do emerging technologies, such as artificial intelligence (AI) and the Internet of Things (IoT), impact job creation and labor reallocation in today's evolving job market?

Initially we are going to discuss AI and IoT with their fruitful working in circular economy and their implication.

4.1.1 Artificial Intelligence (AI)

AI, also known as artificial intelligence, refers to the replication of human intellect in machines, enabling them to accomplish tasks that typically require human intelligence. AI systems are designed to mimic various cognitive functions such as problem-solving, learning, reasoning, perception, and comprehension of language. AI has a wide range of applications across various industries and domains as shown in Figure 4.1. It has the potential to revolutionize industries, enhance decision-making, and improve efficiency across multiple sectors. However, it also poses challenges that need to be addressed, including ethical dilemmas, regulation, and responsible use to ensure its societal benefits (Singh, 2017a, b; Naim & Kautish, 2022; Galgal et al., 2023).

Here are some key aspects of AI:

- Machine Learning and Deep Learning: Machine learning is a subset of AI that involves training algorithms to learn from data and make predictions or decisions based on acquired knowledge (Behera et al., 2023; Embia et al., 2023b). It is a crucial element of AI and encompasses techniques such as supervised learning, unsupervised learning, and reinforcement learning. On the other hand, deep learning is a subset of machine learning that utilizes artificial neural networks inspired by the human brain. Deep neural networks are particularly effective in tasks such as recognizing images, processing natural language, and recognizing speech.
- Natural Language Processing (NLP) and Computer Vision: NLP focuses on enabling machines to understand, interpret, and generate human language. It finds applications in chatbots, language translation, sentiment analysis, and text summarization (Singh, 2019b). Computer vision, on the other hand, involves teaching machines to interpret and understand visual information from images or videos. Applications include image recognition, object detection, and facial recognition. Reinforcement learning, a type of machine learning, is used to enable an agent to learn decision-making

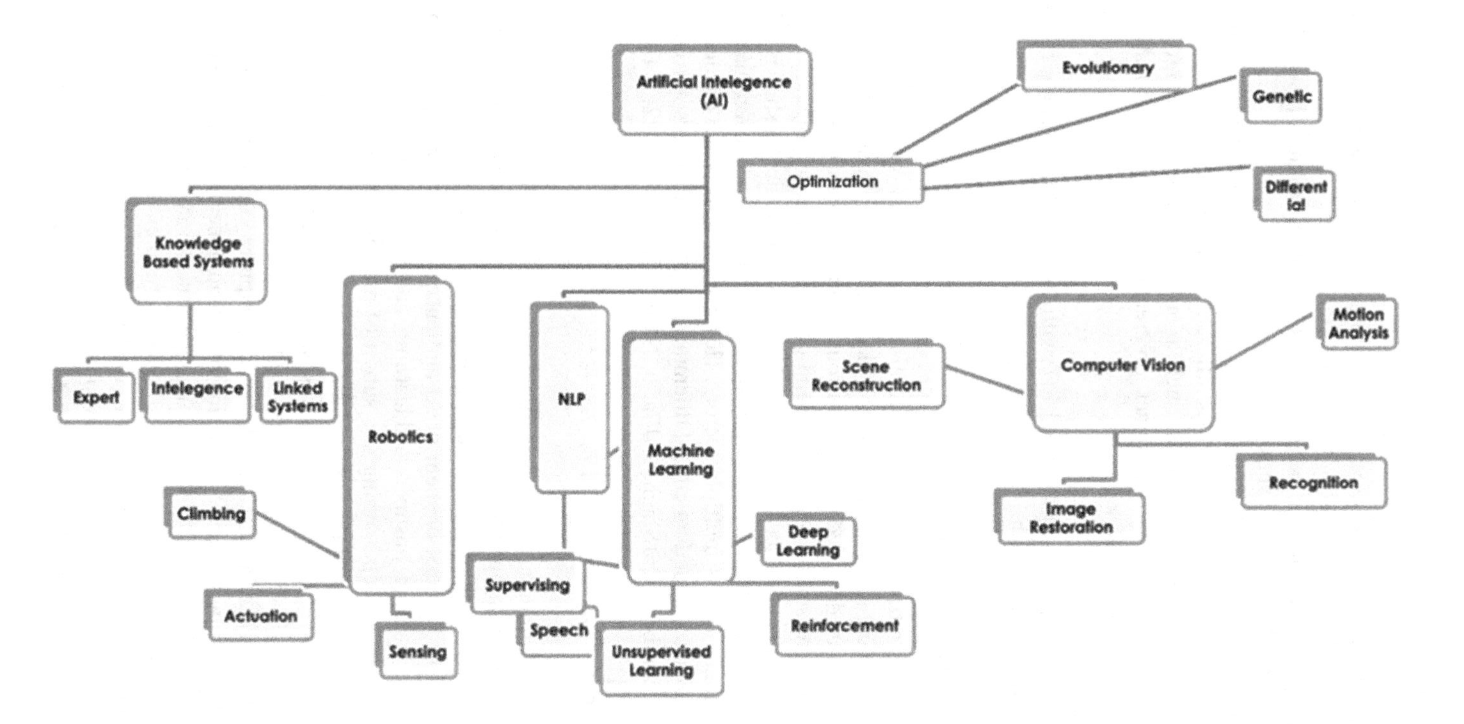

FIGURE 4.1 AI development with different field.

through interaction with an environment. It finds applications in robotics, autonomous vehicles, and game playing.

- AI in Healthcare, Finance, Autonomous Vehicles: AI is used in medical diagnosis, drug discovery, personalized treatment plans, and health monitoring. AI-powered tools can analyze medical images, predict disease outcomes, and assist in surgical procedures. In finance, AI is employed for algorithmic trading, fraud detection, credit scoring, and customer service chatbots. It helps in analyzing large datasets and making financial predictions. Self-driving cars and drones rely on AI technologies, including computer vision and machine learning, for navigation and real-time decision-making.
- AI in Education, Research and Development, Ethical and Social Considerations: AI can personalize learning experiences, automate administrative tasks, and provide adaptive learning platforms to enhance education. As AI becomes more prevalent, ethical concerns related to bias, privacy, transparency, and job displacement have gained prominence. Ensuring responsible AI development and deployment is crucial. AI is a rapidly evolving field, with ongoing research and development in areas like quantum computing, explainable AI, and AI ethics.

4.1.2 Internet of Things (IoT)

The Internet of Things (IoT) refers to the network of interconnected physical objects or "things" embedded with sensors, software, and connectivity, allowing them to collect and exchange data over the internet. These objects can be everyday items like appliances, vehicles, industrial machines, wearable devices, and even infrastructure components like streetlights and traffic sensors. IoT has the potential to revolutionize industries and improve quality of life, but it also requires careful consideration of security and privacy concerns (Singh, 2019a). As IoT technologies continue to evolve, they are likely to become increasingly integrated into various aspects of daily life and industry operations as shown in Figure 4.2. IoT enables these objects to communicate and share data, often in real time, without human intervention. Here are key aspects of IoT (Adams et al., 2018, Chhotaray et al., 2024):

- **Sensors and Data Collection, Connectivity, Data Processing, and Analytics**: IoT devices are equipped with various types of sensors, including temperature, humidity, GPS, motion, and more. These sensors collect data from the physical world. IoT devices use various communication protocols like Wi-Fi, Bluetooth, cellular networks, and others to transmit data to other devices or systems, often through the internet. Collected data is processed, analyzed, and sometimes stored in cloud-based or edge computing systems. Advanced analytics and machine learning can extract valuable insights from the data.
- **Remote Monitoring and Control, Smart Home, and Consumer Applications**: IoT enables remote monitoring and control of devices and systems. For example, you can remotely adjust the temperature of your smart thermostat or monitor security cameras through your smartphone. IoT

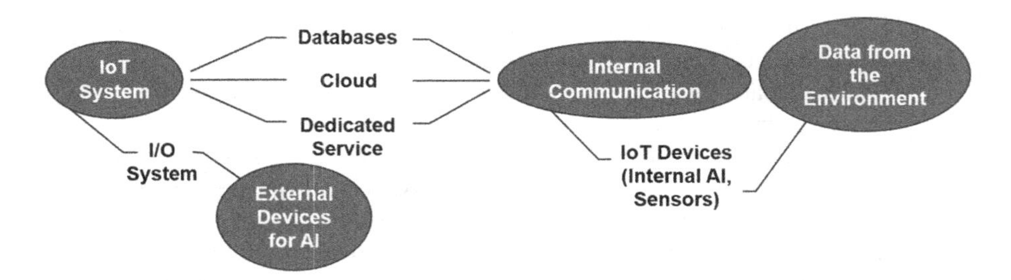

FIGURE 4.2 IoT Development with different working field.

has led to the development of smart homes, where connected devices like smart thermostats, lights, and voice assistants enhance convenience, energy efficiency, and security.

- **Smart Cities, Industrial IoT**: IoT is integral to building smart cities, where connected infrastructure such as smart grids, waste management systems, and traffic control enhances efficiency, sustainability, and quality of life. In industrial settings, IoT devices are used for asset tracking, predictive maintenance, process optimization, and supply chain management, among other applications.
- **Healthcare, Transportation and Logistics, Agriculture**: IoT devices like wearable fitness trackers and remote patient monitoring tools are used to collect health data and improve healthcare outcomes. IoT is used in transportation and logistics for vehicle tracking, route optimization, fleet management, and real-time monitoring of cargo conditions (Singh & Sharma, 2020). Precision agriculture uses IoT to monitor soil conditions, crop health, and irrigation needs, optimizing farming practices and resource usage.
- **Security and Privacy, Scalability**: IoT devices raise concerns about data security and privacy, as they collect and transmit sensitive information. Ensuring robust security measures is essential (Singh, 2017c). IoT networks can scale to accommodate a vast number of devices, making it suitable for both small-scale applications and massive deployments.
- **Environmental Monitoring, Energy Efficiency, Challenges**: IoT plays a crucial role in environmental conservation by monitoring air quality, water quality, and wildlife tracking, among other applications. IoT can help reduce energy consumption by enabling smarter energy management and optimizing power usage in homes, buildings, and industrial facilities. IoT faces challenges related to interoperability, standardization, security, and data privacy. Addressing these challenges is essential for the continued growth of IoT.

4.2 ARTIFICIAL INTELLIGENCE AND THE INTERNET OF THINGS

The combination of AI and the IoT is a powerful synergy that enables the creation of smart, interconnected systems with enhanced capabilities (Ahamad et al., 2021). Here's how AI and IoT work together (see Figure 4.3):

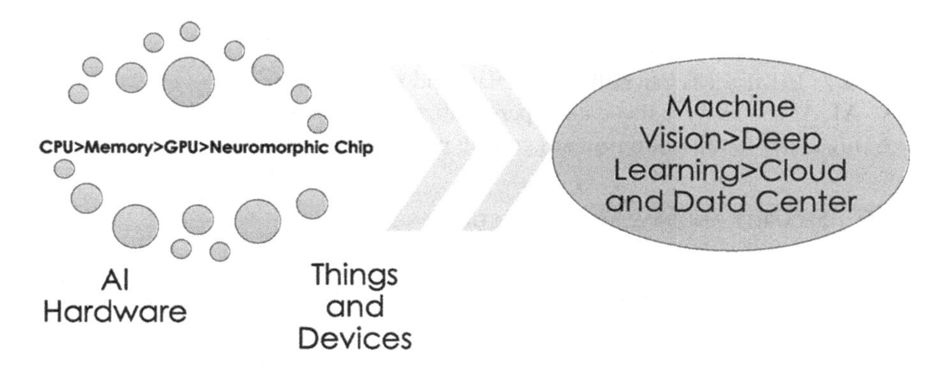

FIGURE 4.3 AI with IoT functionality.

4.2.1 Data Collection and Sensing

- **IoT**: IoT devices equipped with sensors collect data from the physical world. These sensors can measure temperature, humidity, pressure, location, motion, and much more.
- **AI**: AI can process and analyze the vast amounts of data generated by IoT sensors. Machine learning models can recognize patterns, anomalies, and insights in the data.

4.2.2 Data Processing and Analysis

- **IoT**: IoT devices send data to cloud or edge computing platforms for storage and analysis.
- **AI**: AI algorithms, running in the cloud or at the edge, process the data in real-time. They can make predictions, detect trends, and provide actionable insights.

4.2.3 Real-time Decision-Making

- **IoT**: IoT devices can be equipped with AI algorithms to make real-time decisions locally (edge computing) without needing to send data to the cloud.
- **AI**: AI models can make complex decisions based on sensor data, enabling autonomous control and response in IoT systems.

4.2.4 Predictive Maintenance

- **IoT**: IoT sensors monitor the condition of machinery and equipment in real-time.
- **AI**: AI algorithms can predict when equipment is likely to fail, enabling maintenance teams to perform preventive maintenance and avoid costly breakdowns (Embia et al., 2023a).

4.2.5 Personalization and User Experience

- **IoT**: IoT devices can collect user data and preferences.
- **AI**: AI can analyze this data to personalize user experiences, such as adjusting smart home settings or suggesting personalized content.

4.2.6 Security and Anomaly Detection

- **IoT**: IoT devices can detect security events and anomalies.
- **AI**: AI-powered security systems can analyze IoT data to identify potential security threats and respond in real-time.

4.2.7 Energy Efficiency

- **IoT**: IoT can optimize energy usage based on sensor data.
- **AI**: AI algorithms can learn and adapt to patterns of energy consumption to further optimize energy efficiency.

4.2.8 Healthcare and Remote Monitoring

- **IoT**: Medical IoT devices can monitor patient health and vital signs.
- **AI**: AI can analyze this data for early disease detection and alert healthcare professionals or patients.

4.2.9 Traffic Management and Autonomous Vehicles

- **IoT**: Traffic sensors and vehicle sensors generate vast amounts of data.
- **AI**: AI systems can process this data for real-time traffic management and enable autonomous driving.

4.2.10 Smart Cities

- **IoT**: IoT sensors are used to monitor various aspects of urban life, from traffic to waste management.
- **AI**: AI can analyze this data to optimize city services and improve overall efficiency and sustainability.

4.2.11 Environmental Monitoring

- **IoT**: IoT sensors can monitor air quality, water quality, and wildlife.
- **AI**: AI can analyze environmental data to provide insights into conservation efforts and pollution control.

There are several benefits associated with the integration of AI and IoT, including improved decision-making, resource optimization, automation, and enhanced user experiences (Bag et al., 2021). However, it also raises important considerations

TABLE 4.1
Tabular Format Summarizing the Integration of AI with IoT

Aspect	IoT	AI
Data Collection and Sensing	IoT devices equipped with sensors collect data from the physical world.	AI processes and analyzes the data generated by IoT sensors.
Data Processing and Analysis	IoT devices send data to cloud or edge computing platforms for storage and analysis.	AI algorithms process and analyze the data in real time, detecting patterns and insights.
Real-time Decision-Making	IoT devices can make real-time decisions locally (edge computing) based on sensor data.	AI models can make complex real-time decisions based on sensor data.
Predictive Maintenance	IoT sensors monitor equipment condition.	AI predicts equipment failure and enable updates.
Personalization and User Experience	IoT devices collect user data and preferences.	AI analyzes user data to provide personalized experiences and recommendations.
Security and Anomaly Detection	IoT devices detect security events and anomalies.	AI analyzes IoT data to identify security threats and anomalies.
Energy Efficiency	IoT optimizes energy usage based on sensor data.	AI algorithms adapt to patterns of energy consumption to improve efficiency.
Healthcare and Remote Monitoring	Medical IoT devices monitor patient health and vital signs.	AI analyzes medical IoT data for early disease detection and remote patient care.
Traffic Management and Autonomous Vehicles	IoT traffic sensors and vehicle sensors generate data.	AI processes IoT data for real-time traffic management and enables autonomous driving.
Smart Cities	IoT sensors monitor urban aspects like traffic and waste management.	AI analyzes IoT data to optimize city services and enhance urban efficiency.
Environmental Monitoring	IoT sensors monitor air quality, water quality, and wildlife.	AI analyzes environmental data to inform conservation efforts and pollution control.

regarding data privacy, security, and the responsible use of these technologies, which must be addressed as they become more integrated into our daily lives and industries (Belhadi et al., 2021, Embia et al., 2023a).

The integration of AI and IoT enhances decision-making, resource optimization, automation, and user experiences across various domains (Chen et al., 2020). However, it also introduces considerations such as data privacy and security that must be addressed for responsible and effective implementation (Singh, 2017c; Centobelli et al., 2021). The AI and IoT in the circular economy enable real-time monitoring clearly shown in Figure 4.4.

The integration of AI and IoT in the circular economy enables real-time monitoring, optimization of resource usage, waste reduction, and informed decision-making. Together, they contribute to more sustainable and efficient circular practices while addressing environmental and economic challenges (Cheng and Wang, 2021).

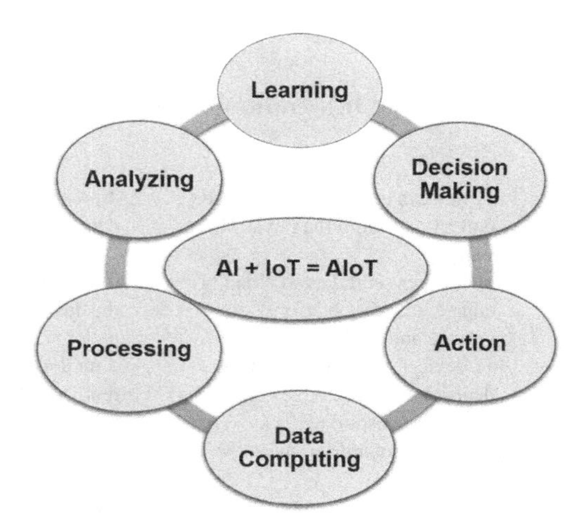

FIGURE 4.4 AI and IoT in the circular economy enable real-time monitoring.

TABLE 4.2
Integration of AI with IoT in the Context of the Circular Economy

Aspect	IoT in Circular Economy	AI in Circular Economy
Data Collection and Sensing	IoT devices equipped with sensors collect data related to resource usage, product condition, and waste flows in real-time. These sensors monitor resource consumption, recycling processes, and waste management.	AI processes and analyzes data generated by IoT sensors to identify patterns and trends related to circular practices.
Real-time Monitoring and Control	IoT enables real-time monitoring and control of resource utilization and product life cycle. It can trigger alerts and actions based on predefined thresholds or conditions.	AI systems provide real-time insights and recommendations for optimizing resource usage and circular processes.
Predictive Maintenance	IoT sensors monitor equipment and machinery, enabling predictive maintenance strategies.	AI predicts equipment failures and maintenance needs, reducing downtime and supporting resource-efficient operations.
Supply Chain Optimization	IoT enhances supply chain visibility by tracking the movement and condition of materials and products throughout the cycle.	AI optimizes supply chains by analyzing large datasets to improve logistics, inventory management, and demand forecasting.
Product Life -Cycle Management	IoT data informs decisions on product redesign, refurbishment, or recycling.	AI analyzes product usage data to inform product life-cycle management strategies, facilitating circular product design.

TABLE 4.2 (CONTINUED)

Aspect	IoT in Circular Economy	AI in Circular Economy
Waste Sorting and Recycling	IoT-based sensors improve waste sorting efficiency, optimizing recycling processes.	AI-driven robotics and automation enhance waste sorting and recycling processes, increasing resource recovery.
Consumer Engagement	IoT provides consumers with information on product origins, sustainability, and circularity, influencing purchasing decisions.	AI personalizes consumer engagement by analyzing data and delivering tailored recommendations for sustainable choices.
Environmental Impact Assessment	IoT data helps assess the environmental impact of circular initiatives, tracking metrics related to waste reduction and resource efficiency.	AI analyzes data to assess the impact of circular economy practices on resource efficiency, waste reduction, and economic benefits, aiding in decision-making.

4.3 SIGNIFICANT ROLE OF AI AND IOT

AI and IoT can play a significant role in promoting circular economy models and evaluating their impact on job creation and reallocation (Grafström & Aasma, 2021). Circular economy models aim to minimize waste, promote sustainability, and extend the lifespan of products and resources (Gupta et al., 2021). Here's how AI and IoT can contribute to this and assess their effects on employment:

- Resource Optimization: Utilizing AI algorithms, data obtained from IoT sensors deployed in manufacturing, logistics, and supply chain operations can be analyzed to optimize the utilization of resources. By minimizing inefficiencies and guaranteeing effective allocation of resources, the principles of a circular economy can be better implemented, resulting in increased employment prospects in resource management and optimization.
- Product Life-Cycle Management: By incorporating IoT sensors into products, real-time data concerning their status and usage can be collected. This data can then be analyzed by AI to optimize maintenance schedules and prolong the lifespan of products. This approach can create and sustain jobs in the maintenance and repair sectors.
- Predictive Maintenance: The implementation of AI-driven predictive maintenance can reduce the necessity for continuous manufacturing and resource extraction by ensuring the longevity of products and equipment. This can indirectly impact job creation by stabilizing employment within the manufacturing industries.
- Supply Chain Optimization: The integration of IoT sensors and AI can enhance the visibility and traceability of supply chains. This not only reduces waste but also creates more job opportunities in logistics, tracking, and inventory management.

- Recycling and Waste Management: By automating sorting and material identification tasks, AI can improve the efficiency of recycling processes. This can result in the generation of jobs within the recycling industry. Furthermore, the utilization of IoT to monitor waste bins and optimize waste collection routes can further enhance resource efficiency and create employment in waste management.
- Circular Business Models: Through the analysis of data on product usage, customer preferences, and market trends, AI can support the development of circular business models. This can drive innovation and lead to the creation of new jobs in the realms of product design, remanufacturing, and services.
- Consumer Engagement: IoT devices and AI-powered applications can provide consumers with information regarding the environmental impact of their choices. This can stimulate demand for sustainable products and services, potentially resulting in job opportunities within these sectors.
- Impact Assessment: By analyzing data from various sources, such as economic indicators and employment data, AI can evaluate the impact of circular economy initiatives on job creation and reallocation. This data can inform policy decisions and investments in sustainable practices.
- Training and Skill Development: The adoption of AI and IoT in circular economy initiatives may necessitate a workforce with specialized skills (Singh, 2017a). This can generate job opportunities in training and skill development programs.
- Monitoring and Compliance: IoT sensors can be deployed to monitor and ensure compliance with circular economy regulations. This can lead to job creation in regulatory oversight and compliance enforcement.

To effectively evaluate the impact on job creation and reallocation (Iqbal et al., 2021), it's crucial to collect and analyze relevant data over time, conduct impact assessments, and collaborate with stakeholders from government, industry, and academia. Furthermore, policies and incentives may be needed to encourage the adoption of circular economy practices and the growth of related job sectors (Kawaguchi, 2019).

The integration of AI and IoT in sustainable development enhances data-driven decision-making (see Figure 4.5), resource optimization, environmental conservation, and resilience in the face of challenges (Khan et al., 2021). Together, they contribute to achieving sustainable development goals and promoting a more sustainable and resilient future (Kurniawan et al., 2021).

4.4 CIRCULAR ECONOMY CONCEPTS

Circular economy concepts are economic systems intended to minimize waste, optimize resource utilization, and foster sustainability (Naim & Malik, 2022; Upadhyay et al., 2021). Unlike the traditional linear economy, which follows a "take-make-dispose" approach, circular economy concepts prioritize continuous product and material reuse, recycling, and remanufacturing (Naim & Malik, 2022). The different challenges, sustainable development goals, and basic characteristics of a circular

TABLE 4.3
Integration of AI with IoT in the Context of Sustainable Development

Aspect	IoT in Sustainable Development	AI in Sustainable Development
Data Collection and Sensing	IoT devices equipped with sensors collect data related to environmental parameters, resource usage, and infrastructure conditions. These sensors monitor air quality, water quality, energy consumption, and more.	AI processes and analyzes data generated by IoT sensors to derive insights and patterns relevant to sustainable development goals.
Real-time Monitoring and Control	IoT enables real-time monitoring of environmental conditions, enabling swift responses to changes or issues.	AI-driven systems provide real-time insights and control mechanisms based on data from IoT devices, optimizing resource usage and infrastructure management.
Environmental Conservation	IoT supports wildlife tracking, habitat monitoring, and environmental conservation. It contributes to biodiversity preservation.	AI analyzes ecological data to support conservation efforts by identifying trends and risks to ecosystems and wildlife.
Resource Efficiency	IoT optimizes resource usage in agriculture, industry, and urban planning.	AI-driven optimization algorithms improve resource efficiency across various sectors.
Renewable Energy Integration	IoT enables efficient management of renewable energy sources, such as solar and wind. It supports grid stability and sustainability.	AI algorithms forecast energy demand and supply, improving renewable energy integration and grid management for sustainable energy.
Water Management	IoT sensors monitor water quality, availability, and distribution in real time. They support sustainable water resource management and conservation.	AI analyzes water data for efficient management, quality control, and distribution to ensure sustainable water resource management and conservation.
Smart Cities and Infrastructure	IoT contributes to the development of smart cities by improving infrastructure efficiency and sustainability.	AI optimizes urban infrastructure by analyzing IoT data to enhance transportation, energy, and waste management in urban areas.
Disaster Management	IoT provides real-time data for early warning systems and disaster response.	AI processes IoT data to predict natural disasters, assess damage, and optimize emergency response strategies.
Public Health	IoT supports health monitoring and disease surveillance, particularly in remote areas. It aids in healthcare access and early intervention.	AI analyzes health data from IoT devices to identify health trends and potential outbreaks.
Agricultural Sustainability	IoT sensors monitor soil conditions, crop health, and weather for sustainable farming.	AI-powered precision agriculture uses IoT data to optimize farming practices, conserve resources, and increase crop yields.
Waste Management	IoT sensors improve waste sorting and recycling processes, reducing waste and supporting circular economy goals.	AI enhances waste management by optimizing recycling, reducing waste generation, and improving waste-to-energy processes.
Eco-friendly Transportation	IoT supports smart transportation systems for efficient and eco-friendly mobility.	AI-powered transportation systems use IoT data for real-time traffic management, autonomous vehicles, and sustainable transportation.

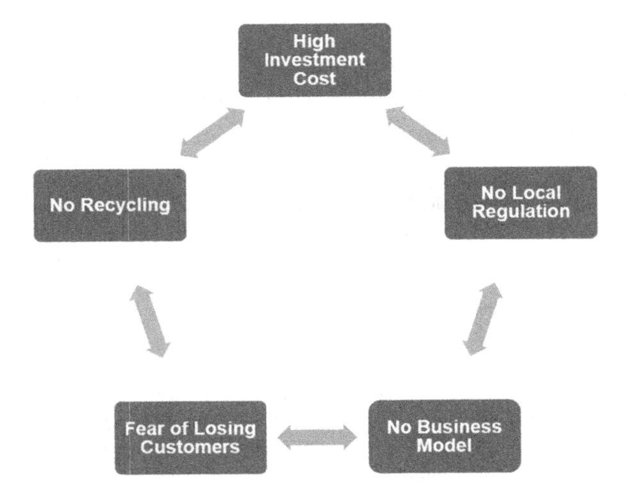

FIGURE 4.5 Drivers of IoT growth.

economy are shown in Figure 4.6–4.8 respectively. These concepts are characterized by a range of fundamental principles and strategies:

- Efficient Resource Usage: Circular economies give priority to resource efficiency, striving to extract maximum value while minimizing waste. This involves reducing material consumption, energy usage, and water consumption in production processes.
- Longevity in Design: Products are designed to have longer lifespans, easy repairability, and standardized components. This extends product life cycles and reduces the need for frequent replacements.
- Reuse: Instead of discarding products after use, circular economies promote reuse. This may involve refurbishing or reselling used products, renting or leasing items, or implementing systems for sharing or exchanging goods.
- Recycling and Remanufacturing: Materials from discarded products are collected and recycled to create new products. Additionally, remanufacturing entails disassembling and reconstructing products to their original specifications, reducing the necessity for new manufacturing.
- Resource Recovery: Circular economies concentrate on recovering and reutilizing valuable materials from waste streams. This can include extracting metals from electronic waste or converting organic waste into compost or biogas.
- Extended Producer Responsibility: Manufacturers bear responsibility for the entire life cycle of their products, including collection, recycling, and disposal. This incentivizes them to design products that are easier to recycle and have reduced environmental impact.
- Digital Technologies: Technologies such as AI and IoT play a pivotal role in monitoring and optimizing circular economy processes. They

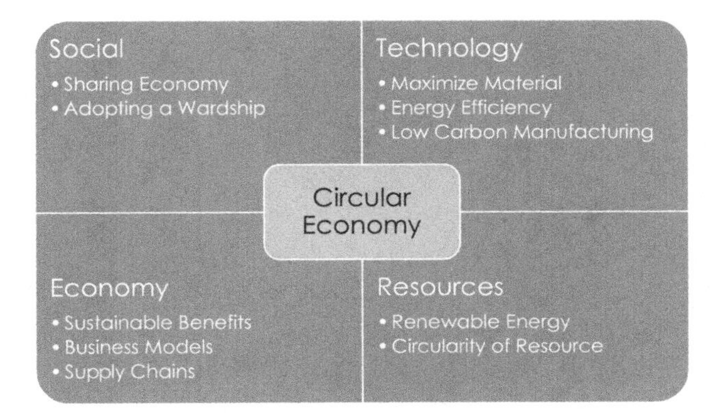

FIGURE 4.6 Challenges in circular economy.

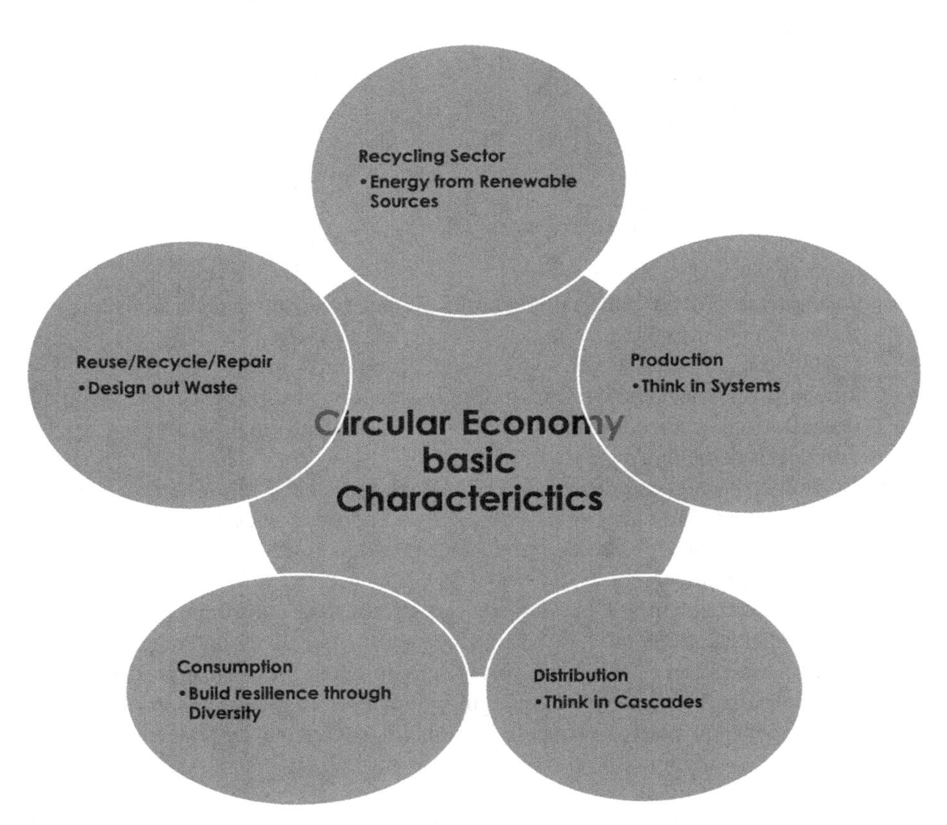

FIGURE 4.7 Sustainable development goals and promoting a more sustainable resilient future.

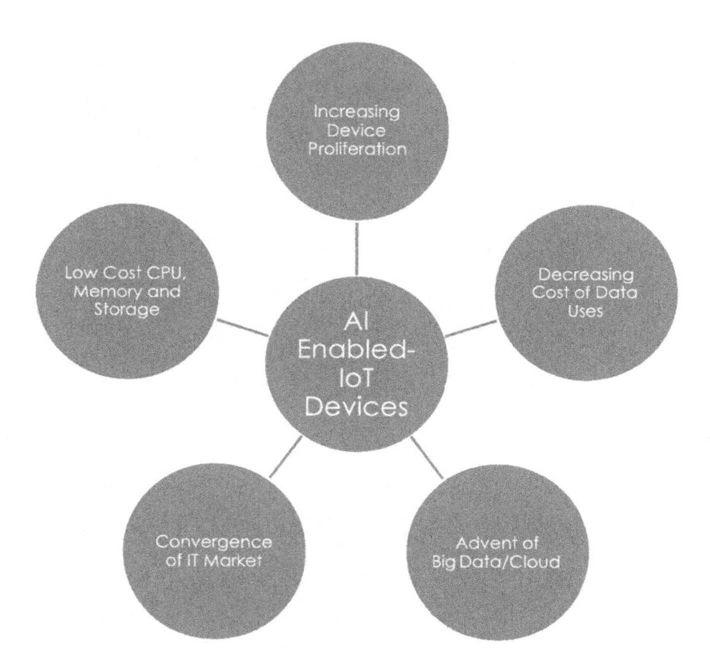

FIGURE 4.8 Five fundamental characteristics.

facilitate improved material tracking, predictive maintenance, and data-driven decision-making.

- Collaboration: Circular economy models frequently involve collaboration among diverse stakeholders, including businesses, governments, non-profit organizations, and consumers. Partnerships and information sharing are crucial for achieving success.
- Consumer Awareness: Educating consumers about the advantages of circular economy practices is essential. Consumers are encouraged to make sustainable choices, such as purchasing products with longer lifespans or recycling properly.
- Policy and Regulation: Governments can significantly contribute to promoting circular economy models through legislation, incentives, and regulations. These policies can establish a favorable environment for circular practices.
- Innovation and Research: Encouraging innovation in materials, production techniques, and business models is vital for advancing circular economy objectives.

Circular economy models offer several benefits, including reduced environmental impact, decreased resource depletion, cost savings, and job creation in sectors related to recycling, repair, and refurbishment (Rehman Khan et al., 2021; Rejeb et al., 2020). By keeping materials and products in use for longer and minimizing waste, circular economies aim to create more sustainable and resilient economic systems (Tiwari & Khan, 2020).

4.5 AI AND IOT IMPACT ON JOB CREATION AND REALLOCATION

AI and IoT technologies have a significant impact on job creation and reallocation across various industries (Roy & Roy, 2019). While these technologies can automate certain tasks and lead to job displacement in some areas, they also create new opportunities and roles in others (Rogers & Katarina, 2021).

4.5.1 Automation and Job Displacement

The utilization of AI-driven automation has the potential to substitute repetitive and mundane tasks, especially in sectors such as manufacturing and data entry. Consequently, this could result in job displacement within these specific industries. The implementation of IoT can automate the monitoring and maintenance processes, thereby reducing the necessity for manual inspections and repairs in domains like utilities and infrastructure. The integration of AI and IoT has the capability to automate repetitive and routine tasks in industries including manufacturing, logistics, and data entry. Although this may lead to job displacement in certain areas, it can also liberate workers to engage in more strategic and innovative roles. As automation technologies progress, it is possible that certain low-skilled positions will become obsolete. Individuals employed in these roles may need to acquire new proficiencies in order to remain employable. The introduction of AI-driven automation can potentially result in the displacement of specific jobs, particularly those that entail repetitive tasks. For instance, occupations within manufacturing, data entry, and basic customer support could be automated. Nevertheless, automation can also generate new employment opportunities in domains such as AI system maintenance and programming. These positions entail proficiencies related to AI development and maintenance, thereby creating prospects for workers equipped with these skills.

4.5.2 Job Enhancement through Augmentation

The implementation of AI and IoT can enhance the capabilities of human workers. In sectors like healthcare, AI has the ability to aid doctors in diagnosing diseases and devising treatment plans, while IoT devices can provide real-time monitoring of patients. Workers can be trained to collaborate with AI and IoT technologies, resulting in job enhancement rather than displacement (Singh, 2017a, b). The expansion of AI and IoT technologies provides employment opportunities in fields such as software development, data analytics, and hardware engineering (Sattarian et al., 2019; Singh & Sharma (2020); Singh 2017c). Proficient professionals are required to design, develop, and maintain these systems. As the adoption of AI and IoT increases, there is a growing demand for experts in AI ethics, policy, and regulation. This encompasses roles in data privacy, cybersecurity, and legal compliance.

4.5.3 New Job Categories

The development and maintenance of AI and IoT systems create new job categories, including AI engineers, data scientists, IoT specialists, and cybersecurity experts.

AI-driven applications in customer service create job opportunities in chatbot development, training, and support.

4.5.4 Data Management and Analysis

AI and IoT generate vast amounts of data. This leads to increased demand for professionals who can manage, analyze, and make decisions based on this data, such as data analysts and data scientists.

4.5.5 Cybersecurity

As more devices and systems become connected through IoT, the need for cybersecurity experts to protect against cyber threats and vulnerabilities grows. This results in job opportunities in cybersecurity and ethical hacking. As IoT devices become more prevalent, the demand for cybersecurity experts to protect against cyber threats and vulnerabilities in these interconnected systems is expected to increase.

4.5.6 Predictive Maintenance

IoT-enabled predictive maintenance roles involve analyzing data from sensors to anticipate when equipment or machinery requires maintenance. This field offers new job opportunities.

4.5.7 AI in Healthcare

AI applications in healthcare, like medical image analysis and drug discovery, create demand for healthcare professionals with AI expertise. IoT devices are transforming healthcare by enabling remote patient monitoring and telehealth services. This creates new roles for healthcare professionals and technicians to manage and interpret patient data.

4.5.8 AI in Education

AI-powered educational tools and personalized learning systems require educators and developers to create, manage, and implement these technologies.

4.5.9 IoT in Agriculture

IoT technologies in agriculture, such as precision farming and smart irrigation, lead to new roles in agricultural technology management. IoT sensors and AI can improve crop management and yield optimization in agriculture. This can lead to job opportunities in precision farming, data analysis, and agricultural technology.

4.5.10 Smart Cities

IoT-driven smart city initiatives create jobs in urban planning, infrastructure management, and data analytics for optimizing city services. IoT technologies play a vital role in creating smart cities and improving infrastructure. This leads to job opportunities in urban planning, IoT system deployment and maintenance, and related fields.

4.5.11 AI-Augmented Workforce

AI and IoT can augment human capabilities, leading to increased productivity in various fields. This may result in job reallocation rather than displacement as workers focus on more value-added tasks. In fields like healthcare, finance, and logistics, AI can assist professionals by providing data-driven insights, which can lead to better decision-making and improved job performance.

4.5.12 Job Creation in AI Training and Data Labeling

AI systems require large datasets for training and validation. Data labeling and curation tasks are often labor-intensive and can create employment opportunities. Teaching AI models and algorithms can create jobs for trainers and educators specializing in AI and machine learning.

4.5.13 AI and IoT in Healthcare

IoT and AI can enable telemedicine services, creating new roles for remote healthcare professionals and support staff. AI can assist radiologists and pathologists in medical image analysis, potentially increasing the efficiency of diagnosis.

4.5.14 IoT in Manufacturing and Maintenance

IoT sensors can predict equipment failures and reduce downtime in manufacturing. This can create jobs in maintenance and repair. IoT can enhance supply chain management, leading to job creation in logistics, inventory management, and transportation.

4.6 IMPLICATION OF AI AND IOT IN CIRCULAR ECONOMY ON JOB CREATION AND REALLOCATION

The effects of AI and IoT on job creation and redistribution within the circular economy are diverse. The incorporation of these technologies into the circular economy can have multiple consequences for the labor market. Skilled professionals are needed to develop, implement, and maintain AI and IoT systems, which in turn create job opportunities in areas such as data analysis, artificial intelligence, management of IoT devices, and consulting on sustainability. To effectively utilize AI and IoT in the circular economy, the workforce may need to acquire new skills, presenting opportunities for training programs and educational providers to meet the demand. By optimizing resource management, supply chains, and waste reduction, AI and IoT can lead to cost savings, thereby creating favorable conditions for business expansion and the potential generation of job opportunities. The combination of AI, IoT, and the circular economy can spark the establishment of startups and new enterprises that focus on sustainability, driving job growth. Rather than resulting in job losses,

AI and IoT may instead reallocate tasks within industries. This reallocation allows workers to focus on higher-value tasks while routine and manual jobs are automated. In the field of sustainability consulting and management within the circular economy, expertise in sustainability management and consulting is required, leading to job creation in these specialized areas. Governments play a critical role by regulating and incentivizing sustainable practices, which can foster the growth of green industries and subsequently create jobs in these sectors. Given that sustainability is a global concern, the implications of AI and IoT in the circular economy extend beyond a single region. International cooperation and competition can shape job markets worldwide. The implementation of AI and IoT in the circular economy can result in improved environmental outcomes, potentially stimulating job growth in sectors related to environmental protection and conservation. It is crucial to ethically and responsibly utilize AI and IoT while considering ethical considerations in the circular economy to ensure that the positive impacts on job creation outweigh any potential negative effects. The implications of AI and IoT in the circular economy on job creation and redistribution are influenced by various factors, including the rate of technology adoption, government policies, and the adaptability of the workforce to new skill requirements. A well-planned approach that takes these implications into account can maximize the benefits of these technologies on job markets while promoting sustainable practices.

4.7 CONCLUSIONS

The influence of AI and IoT on job generation and redistribution varies across industries and regions. While some job positions may be automated or replaced, there are also ample chances for novel and evolving roles that utilize these technologies. The ability to acquire new skills and adaptability will be crucial for the workforce to succeed in this evolving environment. AI and IoT have the potential to significantly impact job generation and redistribution in various sectors. While AI and IoT may disrupt certain job positions, they also offer significant prospects for job generation, skill development, and career advancement in emerging technology fields and those that embrace new technological trends. To maximize the benefits of these technologies on job markets, investments in education and training are vital to ensure that the workforce is prepared for the changing job landscape. The implementation of AI and IoT technologies can have a substantial effect on job generation and redistribution across multiple industries and sectors. In conclusion, although AI and IoT can result in the automation and replacement of specific jobs, they also create new employment opportunities in technology development, data analysis, and the management of AI and IoT systems. The impact on job generation and redistribution will largely depend on the industry, the rate of technology adoption, and the ability of the workforce to acquire new skills and adapt to changing job roles. Therefore, a focus on education, upskilling, and reskilling will be essential to facilitate workers' transition into the jobs of the future.

REFERENCES

Adams, R., Kewell, B., & Parry, G. (2018). Blockchain for Good? Digital Ledger Technology and Sustainable Development Goals. In *Handbook of Sustainability and Social Science Research. World Sustainability Series*, eds Leal Filho, W., Marans, R., Callewaert, J. (Cham: Springer).

Ahamad, S., Gupta, P., Bikash, A. P., Padma, K. K., Khan, Z., & Faez, H. M. (2021). The role of block chain technology and Internet of Things (IoT) to protect financial transactions in crypto currency market. *Mater. Today Proc.* https://doi.org/10.1016/j.matpr.2021.11.405

Bag, S., Pretorius, J. H. C., Gupta, S., & Dwivedi, Y. K. (2021). Role of institutional pressures and resources in the adoption of big data analytics powered artificial intelligence, sustainable manufacturing practices and circular economy capabilities. *Technol. Forecast. Soc. Change* 163:120420. https://doi.org/10.1016/j.techfore.2020.120420

Behera, B. C., Moharana, B. R., Rout, M., & Debnath, K. (2023). Application of Machine Learning in the Machining Processes: Future Perspective Towards Industry 4.0. *Intelligent Manufacturing Management Systems: Operational Applications of Evolutionary Digital Technologies in Mechanical and Industrial Engineering*, 141–156.

Belhadi, A., Kamble, S. S., Chiappetta, J. C., Mani, V., Rehman Khan, S. A., & Touriki, F. E. (2021). A self-assessment tool for evaluating the integration of circular economy and industry 4.0 principles in closed-loop supply chains. *Int. J. Prod. Econ.* 245:108372. https://doi.org/10.1016/j.ijpe.2021.108372

Centobelli, P., Cerchione, R., Vecchio, P., Del, O. E., & Secundo, G. (2021). Blockchain technology for bridging trust, traceability and transparency in circular supply chain. *Inf. Manag.* 103508. https://doi.org/10.1016/j.im.2021.103508

Chen, T. L., Kim, H., Pan, S. Y., Tseng, P. C., Lin, Y. P., & Chiang, P. C. (2020). Implementation of green chemistry principles in circular economy system towards sustainable development goals: challenges and perspectives. *Sci. Total Environ.* 716:136998. https://doi.org/10.1016/J.SCITOTENV.2020.136998

Cheng, F., & Wang, Y. (2021). Research and application of 3D visualization and Internet of Things technology in urban land use efficiency management. *Displays* 69:102050. https://doi.org/10.1016/j.displa.2021.102050

Chhotaray, P., Behera, B. C., Moharana, B. R., Muduli, K., & Sephyrin, F. T. R. (2024). Enhancement of Manufacturing Sector Performance with the Application of Industrial Internet of Things (IIoT). In *Smart Technologies for Improved Performance of Manufacturing Systems and Services* (pp. 1–19). CRC Press.

Embia, G., Mohamed, A., Moharana, B. R., & Muduli, K. (2023a). Edge Computing-Based Conditional Monitoring. *Intelligent Manufacturing Management Systems: Operational Applications of Evolutionary Digital Technologies in Mechanical and Industrial Engineering*, 249–270.

Embia, G. J., Moharana, B. R., Behera, B. C., Mohmaed, N. H., Biswal, D. K., & Muduli, K. (2023b). Reliability Prediction Using Machine Learning Approach. In *Smart Technologies for Improved Performance of Manufacturing Systems and Services* (pp. 21–37). CRC Press.

Galgal, K. N., Ray, M., Moharana, B. R., Behera, B. C., & Muduli, K. (2023). Quality Control in the Era of IoT and Automation in the Context of Developing Nations. In *Smart Technologies for Improved Performance of Manufacturing Systems and Services* (pp. 39–50). CRC Press.

Grafström, J., & Aasma, S. (2021). Breaking circular economy barriers. *J. Clean. Prod.* 292:126002. https://doi.org/10.1016/j.jclepro.2021.126002

Gupta, H., Kumar, A., & Wasan, P. (2021). Industry 4.0, cleaner production and circular economy: an integrative framework for evaluating ethical and sustainable business performance of manufacturing organizations. *J. Clean. Prod.* 295:126253.

Iqbal, W., Tang, Y. M., Lijun, M., Chau, K. Y., Xuan, W., & Fatima, A. (2021). Energy policy paradox on environmental performance: the moderating role of renewable energy patents. *J. Environ. Manage.* 297:113230. https://doi.org/10.1016/j.jenvman.2021.113230

Kawaguchi, N. (2019). Application of Blockchain to supply chain: flexible Blockchain technology. *Procedia Comput. Sci.* 164:143–148. https://doi.org/10.1016/j.procs.2019.12.166

Khan, S. A. R., Ponce, P., Tanveer, M., Aguirre-Padilla, N., Mahmood, H., & Shah, S. A. A. (2021). Technological innovation and circular economy practices: business strategies to mitigate the effects of COVID-19. *Sustainability* 13:8479. https://doi.org/10.3390/SU13158479

Kurniawan, T. A., Lo, W., Singh, D., Othman, M. H. D., Avtar, R., Hwang, G. H., et al. (2021). A societal transition of MSW management in Xiamen (China) toward a circular economy through integrated waste recycling and technological digitization. *Environ. Pollut.* 277:116741. https://doi.org/10.1016/j.envpol.2021.116741

Naim, A., & Kautish, S. K. (Eds.). (2022). *Building a Brand Image through Electronic Customer Relationship Management*. IGI Global. https://doi.org/10.4018/978-1-6684-5386-5

Naim, A., & Malik, P. K. (2022). *Competitive Trends and Technologies in Business Management*, Nova Science Publisher, USA, February 15, 2022. https://doi.org/10.52305/VIXO9830. ISBN: 978-1-68507-612-2

Rehman Khan S. A., Yu, Z., Sarwat, S., Godil, D. I., Amin, S., & Shujaat, S. (2021). The role of block chain technology in circular economy practices to improve organisational performance. *Int. J. Logist. Res. Appl* https://doi.org/10.1080/13675567.2021.1872512

Rejeb, A., Simske, S., Rejeb, K., Treiblmaier, H., & Zailani, S. (2020). Internet of Things research in supply chain management and logistics: a bibliometric analysis. *Internet Things* 12:100318. https://doi.org/10.1016/J.IOT.2020.100318

Rogers, S., & Katarina, Z. (2021). Big Data-driven algorithmic governance in sustainable smart manufacturing: robotic process and cognitive automation technologies. *Anal. Metaphys* 20:130. https://doi.org/10.22381/am2020219

Roy, M., & Roy, A. (2019). Nexus of Internet of Things (IoT) and Big Data: roadmap for Smart Management Systems (SMgS). *IEEE Eng. Manag. Rev.* 47:53–65. https://doi.org/10.1109/EMR.2019.2915961

Sattarian, M., Rezazadeh, J., Farahbakhsh, R., & Bagheri, A. (2019). Indoor navigation systems based on data mining techniques in internet of things: a survey. *Wirel. Networks* 25:1385–1402. https://doi.org/10.1007/s11276-018-1766-4

Singh, A. K. (2017a). Security and Management in Network: Security of Network Management versus Management of Network Security (SNM Vs MNS). *International Journal of Computer Science and Network Security (IJCSNS)* 17(5):166–173, May-2017, indexed in ISI Thomson Reuter (web of science -ESCI). ISSN: 1738-7906.

Singh, A. K. (2017b). The active impact of human computer interaction (HCI) on economic, cultural and social life. *IIOAB Journal* 8(2):141–146, June-2017, indexed in SCOPUS/ISI Q4. ISSN: 0976-3104

Singh, A. K. (2017c). Persona of social networking in computing and informatics era. *International Journal of Computer Science and Network Security (IJCSNS)* 17(4):95–101, April-2017, indexed in ISI Thomson Reuter (web of science -ESCI), ISSN: 1738-7906.

Singh, A. K. (2019a). An intelligent reallocation of load for cluster cloud environment. *Int. J. Innov. Technol. Expl. Eng. (IJITEE)* 8(8):711–714, indexed in SCOPUS.

Singh, A. K. (2019b). Texture-based real-time character extraction and recognition in natural images. *Int. J. Innov. Technol. Expl. Eng. (IJITEE)* 8(8):3302–3306, indexed in SCOPUS

Singh, A. K., & Sharma, S. D. (2020). Digital era in the Kingdom of Saudi Arabia: Novel strategies of the telecom service providers companies. *Webology* 17(1):227–245, June-2020 indexed in Scopus (Scopus/ISI Q3), https://doi.org/10.14704/WEB/V17I1/a219. ISSN: 1735-188X.

Tiwari, K., & Khan, M. S. (2020). Sustainability accounting and reporting in the industry 4.0. *J. Clean. Prod.* 258:120783. https://doi.org/10.1016/J.JCLEPRO.2020.120783

Upadhyay, A., Mukhuty, S., Kumar, V., & Kazancoglu, Y. (2021). Blockchain technology and the circular economy: implications for sustainability and social responsibility. *J. Clean. Prod.* 293:126130. https://doi.org/10.1016/j.jclepro.2021.126130

5 An EOQ Ordering Policy for Items with Varying Deterioration, Exponential Declining Demand and Shortages

Trailokyanath Singh and Sephali Mohanty
C. V. Raman Global University, Bhubaneswar, India

5.1 INTRODUCTION

In a circular economy, time and cost are two important competitive factors to run businesses. In real-life situations, the economic activity of the EOQ (economic order quantity) inventory system minimizes the waste items as well as maximizes usability of the resources. In a business scenario, the industries or the big companies have been trying to maximize their profit by minimizing time and different system costs on ordering, carrying, deterioration, shortages, and so on. In our day-to-day life, the decaying effect on physical items is a natural process and occurs due to damage, spoilage, evaporation, and so forth. Vegetables, crops, fruits, agricultural products, medicines, drugs, pharmaceuticals, and spare parts are examples of such items. Therefore, the physical loss of items due to the process of deterioration must be taken into consideration while formulating an optimal inventory model.

Definition 1

Deteriorating items are defined as those items which cannot be used for their original purposes. Blood-bank blood, fashion goods, foods, pharmaceuticals, volatile liquids, radioactive substances, electronics components, etc. are some examples of deteriorating items.

Many researchers studied various inventory problems and their solutions for deteriorating items from time to time. Firstly, Ghare and Schrader (1963) formulated the inventory model for physical items in which products or items deteriorate by the negative exponential function of time. Then, Covert and Philip (1973) generalized Ghare and Schrader's (1963) model with suitable deterioration rate. The constant deterioration function was replaced by the two-parameter Weibull density function in their model. In the present scenario, demand for any item always exists in vibrant

DOI: 10.1201/9781003349877-5

state. Therefore, the inclusion of the constant demand in the inventory system rate is not always suitable for the development of models. Later, numerous inventory policies for deteriorating items were studied not only on the assumptions of the constant demand rate but also the basis of dynamic demand patterns such as linearly increasing or decreasing demand, exponential demand, quadratic demand, and others. Hollier and Mak (1983) proposed an inventory policy with different replenishment intervals by incorporating exponential declining demand pattern. Wee (1995) and Jalan et al. (1996) studied inventory models with the incorporation of quadratic demand pattern. Su et al. (1999) introduced an optimal production inventory model with the assumption of shortages and exponentially declining demand pattern. Ghosh and Chaudhuri (2006) established a policy with the factors demand and deterioration as quadratic and constant function of time, respectively. In case of backlogged shortages, the backlogging rate is a representation of waiting time, and the willingness of a customer to wait during the partially backlogged shortages period is reciprocal to the length of delaying or waiting time of the system. In this context, the deteriorating model with time-dependent demands and partial backlogged shortages was developed by Chang and Dye (1999). Ouyang et al. (2005) studied the backlogged inventory system with exponentially declining demand in which the backlogging rate and delaying time are reciprocal to each other. Basu et al. (2006) studied a multi-item inventory model with the factor exponential declining demand and without shortages under the circumstances delay in payments. Singh and Pattanayak (2012) introduced an EOQ optimal ordering policy with varying the time-dependent exponentially declining demand and several delay in payments conditions. The deterioration rate was considered as constant for the formulation of two different cases on permissible delay period in their model. The comprehensive literature review on decaying inventory systems have been published in the papers of Raafat (1991), Li et al. (2010), and Bakker et al. (2012).

The case of partially backlogged shortages happens when newly launched fashionable and cosmetic items and high-tech products like desktops, laptops, android mobiles, and others are launched into the market. Recently, Singh et al. (2021) studied the Weibull distributed deteriorated inventory system with quadratic demand pattern and shortages. The varying deterioration rate and salvage value of items are also incorporated in their model. Some models were developed with the consideration of time-dependent holding cost. In this regard, Swain et al. (2021) introduced the model with both generalized demand pattern and deterioration. Singh et al. (2021) developed an ordering policy for decaying items with dynamic factors such as trapezoidal-type demand, variable deterioration and shortages. Lakshmi and Pandian (2021) developed a production inventory system with exponential inclining demand and varying deterioration. Mohanty and Singh (2023) established an EOQ optimal ordering inventory system with consideration of the cubic demand rate, variable deterioration rate, and the concept of delay in payments. The generalized demand and deterioration is taken as the cubic and linear function of time, respectively. An inventory system varying with several factors such as generalized exponential increasing demand, variable deterioration, and time-dependent holding cost was studied by Mohanty et al. (2023). Generally speaking, the demand of items may vary with respect to the passage of time and constant demand, linearly increasing or linearly decreasing demand and quadratic demand rate, and so forth do not specifically show the demand of certain items such as blood-bank blood, pharmaceuticals, volatile liquids, radioactive substances, newly launched mobile phones, automobiles and their spare parts, and so on. However, the demand pattern of those items might be satisfied by the application of exponential declining demand pattern. Really, the demand for an item varies over time because of the continuous introduction of new items, change in awareness about the item, loss of appeal, and so on. The demand for new items like android mobiles, new models of laptops, and others cannot be continuously sustained when newer and more efficient products are coming to the market.

Nowadays, in order to develop circular economy capabilities, developing countries like India, South Africa, and others focus on quality production aiming at sustainable development features. They have been connecting with the prospective of Industry 4.0 scientific innovation to nurture their economy and strengthen people of their countries. In this context, Mohanty et al. (2022) introduced an optimal replenishment inventory policy with varying deterioration and an exponential declining demand rate. An EOQ optimal inventory model with varying deterioration, a quadratic demand rate, salvage values of items and partially backlogged shortages was studied by Singh et al. (2018). The key resources like human resources, production systems, project management, modern technology, and so on, are also essential for Industry 4.0 execution, sustainable manufacture and growing economy. The effects of several possible resources for industry 4.0 execution, sustainable production, development of a circular economy, and others were studied by Bag et al. (2021).

Generally, there are some items in which deterioration does not take place while stored on shelves, although they start deteriorating after some time. For such cases, time-proportional deterioration represents the time to the process of deterioration. The objective for developing an optimal policy is to formulate a generalized model considering the variable deterioration rate, time-varying exponentially declining demand rate, and shortages. The considered shortages in this system are partially backlogged shortages and its rate and delay time for the next order are reciprocal to each other. The primary goal of the present study is to optimize two lengths of delaying time and length of cycle along with the total cost function of the system. The effectiveness of the model is discussed with consideration of a couple of numerical examples. At the end, the effects on changes of parameters have been discussed with the discussion of sensitivity analysis.

5.2 FUNDAMENTAL ASSUMPTIONS AND NOTATIONS

5.2.1 Fundamental Assumptions

The fundamental assumptions described for the mathematical expressions of the system are as follows:

1. The system needs only one type of item for the formulation.
2. Replenishment of items occurs at the rate infinite.
3. Deterioration acts as variable and is a linear increasing function of time.
4. Demand pattern is known and is expressed as an exponentially declining function of time.

5. Neither replacement nor repairing of units during the prescribed period of the inventory system under consideration.
6. The backlogging rate at any time is defined for negative inventory and is dependent on the next replenishment.

5.2.2 Notations

The following notations are taken for representing the components of the system mathematically.

$\theta_0(t) = \alpha t,\ 0 < \alpha << 1$: the variable deterioration rate.

$D(t) = \begin{cases} Ke^{-\gamma t}, I(t) > 0 \\ D_0,\ \ I(t) \le 0 \end{cases}$: represents the exponentially declining demand rate.

Here, $K(>0)$ & $\gamma(0 < \gamma << 1)$ denote the initial and declining rate of demand, respectively.

$B_0(t) = \dfrac{1}{1+\delta_0(T-t)}$: the backlogging rate (negative inventory). Here, $\delta_0(>0)$ denotes the backlogging parameter which is a constant.

c_1, c_2, c_o, c_s & c_l: cost of holding stock \$/unit time, deterioration \$/unit time, replenishment of item \$/order, shortages \$/unit/unit time and opportunity because of lost sales \$/unit, respectively.

t_s & T: time of the shortages point and cycle length of the system, respectively.

W_{max}, S_{max} & I_0: highest level for cycle, the maximum quantity of backlogged demand and the EOQ for system, respectively.

$I(t)$: inventory level at time t.

$AT(t_s, T)$: minimum total average cost.

5.3 THE MATHEMATICAL FORMULATION

In the beginning, the maximum level of inventory is W_{max} units of items at the initial time $t = 0$. During the period $[0, t_s]$, the present level of stock declines due to the time-proportional deterioration rate and the desired exponentially declining demand rate. At subsequent time t_s, the inventory level achieves zero; after that shortages are recorded in the system (see Figure 5.1).

The differential equation representations of the inventory level in the period $0 \le t \le T$ can be stated as

$$\frac{dI(t)}{dt} + \theta_0(t)I(t) + D(t) = 0, \qquad 0 \le t \le t_s \tag{5.1}$$

where $\theta_0(t) = \alpha t$ (variable deterioration) and $D(t) = Ke^{-\gamma t}$ (exponentially inclined demand rate);

and

$$\frac{dI(t)}{dt} + D(t) \times B_0(t) = 0, \qquad t_s \le t \le T \tag{5.2}$$

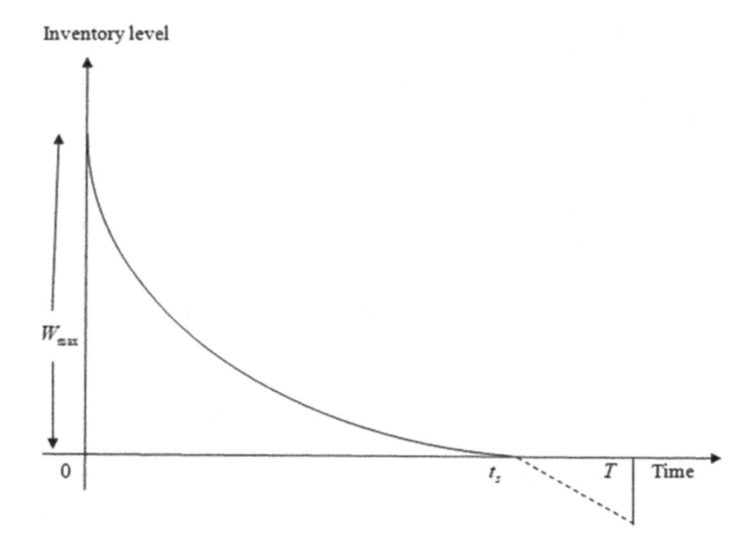

FIGURE 5.1 Inventory-time graph.

where

$$B_0(t) = \frac{1}{1+\delta_0(T-t)}, \ \delta_0 > 0.$$

From Eq. (5.1), integrating factor (denoted by I. F.) is

$$I.F. = e^{\int(\alpha t)dt} = e^{\frac{\alpha t^2}{2}} \tag{5.3}$$

and its solution is

$$I(t).e^{\frac{\alpha t^2}{2}} = -K\int\left(e^{-\gamma t}\right)e^{\frac{\alpha t^2}{2}}dt + A,$$

where A is the constant of integration.

Neglecting the terms containing the factors like $\alpha^2, \alpha^3, \cdots; \gamma^2, \gamma^3, \cdots; \alpha\gamma, \alpha^2\gamma^2, \cdots;$ using the boundary condition $I(t_s) = 0$, the solution of Eq. (5.1) is given by

$$I(t) = K\left[t_s - \frac{\gamma t_s^2}{2} + \frac{\alpha t_s^3}{6} - \left(t - \frac{\gamma t^2}{2} + \frac{\alpha t^3}{6}\right)\right]e^{\left(-\frac{\alpha t^2}{2}\right)}, \qquad 0 \le t \le t_s \tag{5.4}$$

The maximum inventory is obtained by substituting the condition $I(0) = W_{max}$ and is given by

$$I(0) = W_{max} = K\left(t_s - \frac{\gamma t_s^2}{2} + \frac{\alpha t_s^3}{6}\right) \tag{5.5}$$

The solution of the Eq. (5.2) with the condition $I(t_s) = 0$ is calculated by

$$I(t) = \frac{D_0}{\delta_0} \times \left[\ln\left[1+\delta_0 (T-t)\right] - \ln\left[1+\delta_0 (T-t_s)\right] \right], \quad t_s \le t \le T \tag{5.6}$$

The maximum backlogged demand per cycle is obtained computed from Eq. (5.6) by setting $S_{\max} = -I(T)$. Thus,

$$S_{\max} = -I(T) = \frac{D_0}{\delta_0} \times \left[\ln\left[1+\delta_0 (T-t)\right] \right] \tag{5.7}$$

Thus, the EOQ per cycle is given by

$$I_0 = W_{\max} + S_{\max} = K\left(t_s - \frac{\gamma t_s^2}{2} + \frac{\alpha t_s^3}{6} \right) + \frac{D_0}{\delta_0}\left[\ln\left[1+\delta_0 (T-t_s)\right] \right] \tag{5.8}$$

The system cost function equals the ratio of (the sum of costs from holding, deterioration, replenishment, shortages, and opportunity because of lost sales) and (the cycle length).

(i) Cost of holding the inventory (CI):

$$\begin{aligned} CI &= c_1 \int_0^{t_s} I(t)dt = c_1 K \int_0^{t_s} \left[t_s - \frac{\gamma t_s^2}{2} + \frac{\alpha t_s^3}{6} - \left(t - \frac{\gamma t^2}{2} + \frac{\alpha t^3}{6} \right) \right] e^{\frac{\alpha t^2}{2}} dt \\ &= c_1 K \int_0^{t_s} \left[t_s - \frac{\gamma t_s^2}{2} + \frac{\alpha t_s^3}{6} - \left(t - \frac{\gamma t^2}{2} + \frac{\alpha t^3}{6} \right) - \frac{\alpha t_s t^2}{2} + \frac{\alpha t^3}{2} \right] dt \\ &= c_1 K \left[\frac{t_s^2}{2} - \frac{\gamma t_s^3}{3} + \frac{\alpha t_s^4}{12} \right] \end{aligned} \tag{5.9}$$

(by ignoring terms containing $\alpha^2, \alpha^3, \ldots; \gamma^2, \gamma^3, \ldots; \alpha\gamma, \alpha^2\gamma^2, \ldots;$ as $0 < \alpha << 1$ & $0 < \gamma << 1$).

(ii) Cost of deterioration (CW):

$$\begin{aligned} CW &= c_2 \left[W - \int_0^{t_s} D(t)dt \right] = c_2 \left[W - \int_0^{t_s} \left(Ke^{-\gamma t} \right) dt \right] \\ &= c_2 K \left[t_s - \frac{\gamma t_s^2}{2} + \frac{\alpha t_s^3}{6} + \frac{1}{\gamma}\left(e^{-\gamma t_s} - 1 \right) \right] \end{aligned} \tag{5.10}$$

(iii) Cost of replenishment (CR):

$$CR = c_o. \tag{5.11}$$

During the shortage period, there are two cases that need to be determined such as (1) the shortages cost for items and (2) the opportunity cost because of lost sales.

(iv) Cost of shortages (CB):

$$CB = c_s\left[-\int_{t_s}^{T} I(t)dt\right] = \frac{c_s D_0}{\delta_0}\left[T - t_s - \frac{1}{\delta_0}\ln\left[1+\delta_0\left(T-t_s\right)\right]\right] \tag{5.12}$$

(v) Cost of opportunity because of lost sales (CS):

$$\begin{aligned} CS &= c_l D_0 \int_{t_s}^{T}\left[1 - \frac{1}{1+\delta_0\left(T-t\right)}\right]dt \\ &= c_l D_0\left[T - t_s - \frac{1}{\delta_0}\times\ln\left[1+\delta_0\left(T-t_s\right)\right]\right] \end{aligned} \tag{5.13}$$

Therefore, $AT(t_s, T)$, the system cost function per unit time is

$$\begin{aligned} AT\left(t_s,T\right) &= \frac{1}{T}\left[CI + CW + CR + CB + CS\right] \\ &= \frac{K}{T}\left[c_1\left(\frac{t_s^2}{2} - \frac{\gamma t_s^3}{3} + \frac{\alpha t_s^4}{12}\right) + c_2\left[t_s - \frac{\gamma t_s^2}{2} + \frac{\alpha t_s^3}{6} + \frac{1}{\gamma}\left(e^{-\gamma t_s} - 1\right)\right]\right] + \frac{c_o}{T} \\ &\quad + \frac{D_0\left(c_s + \delta_0 c_l\right)}{\delta_0 T}\left[T - t_s - \frac{1}{\delta_0}\ln\left[1+\delta_0\left(T-t_s\right)\right]\right] \end{aligned} \tag{5.14}$$

Thus, the parameters such as shortages time and cycle length, and average cost function are obtained from the system of equations

$$\frac{\partial AT\left(t_s,T\right)}{\partial t_s} = 0 \tag{5.15}$$

and

$$\frac{\partial AT\left(t_s,T\right)}{\partial T} = 0 \tag{5.16}$$

provided that $\frac{\partial^2 AT(t_s,T)}{\partial t_s^2} > 0$, $\frac{\partial^2 AT(t_s,T)}{\partial T^2} > 0$ and

$$\left(\frac{\partial^2 AT(t_s,T)}{\partial t_s^2}\right) \times \left(\frac{\partial^2 AT(t_s,T)}{\partial T^2}\right) - \left(\frac{\partial^2 AT(t_s,T)}{\partial t_s \partial T^2}\right)^2 > 0$$

(see Appendices A, B, and C).

From Eq. (5.15), we have

$$\frac{\partial AT(t_s,T)}{\partial t_s} = \frac{K}{T}\left[c_1\left(t_s - \gamma t_s^2 + \frac{\alpha t_s^3}{3}\right) + c_2\left(1 - \gamma t_s + \frac{\alpha t_s^2}{2} - e^{-\gamma t_s}\right)\right] - \frac{1}{T}\left[\frac{D_0(c_s + \delta_0 c_l)(T - t_s)}{1 + \delta_0(T - t_s)}\right] = 0 \tag{5.17}$$

and from equation (5.16), we obtain

$$\begin{aligned}\frac{\partial AT(t_s,T)}{\partial T} &= \frac{1}{T}\left[\frac{D_0(c_s + \delta_0 c_l)(T - t_s)}{1 + \delta_0(T - t_s)}\right] \\ &- \frac{K}{T^2}\left[c_1\left(\frac{t_s^2}{2} - \frac{\gamma t_s^3}{3} + \frac{\alpha t_s^4}{12}\right) + c_2\left[t_s - \frac{\gamma t_s^2}{2} + \frac{\alpha t_s^3}{6} + \frac{1}{\gamma}\left(e^{-\gamma t_s} - 1\right)\right]\right] \\ &- \frac{c_o}{T^2} - \frac{D_0(c_s + \delta_0 c_l)}{\delta_0 T^2}\left[T - t_s - \frac{1}{\delta_0}\ln\left[1 + \delta_0(T - t_s)\right]\right] = 0\end{aligned} \tag{5.18}$$

Firstly, $t_s^* \,\&\, T^*$, the optimal times, are computed from both Eqs. (5.17) and (5.18). Thereafter, the economic order quantity (I_0^*) and the minimum average cost function per unit time ($AT^*(t_s, T)$) can be calculated by putting the obtained values of $t_s^* \,\&\, T^*$ in the Eqs. (5.8) and (5.14), respectively.

5.4 NUMERICAL ILLUSTRATIONS

The proposed model is described for the case of backlogged shortages with the following data sets:

EXAMPLE 5.1

Let the parameters be assigned with their proper units: $\alpha = 0.004$, $K = 16$, $\gamma = 0.02$, $D_0 = 8$, $\delta_0 = 2$, $c_1 = 0.6$, $c_2 = 1.6$, $c_o = 10$, $c_s = 2.4$, and $c_l = 2$. Solving Eqs. (5.17) and (5.18), we get $t_s^* = 1.29141$ unit time and $T^* = 1.74533$ unit time. The EOQ and the optimal cost function are $I_0^* = 23.0026$ units and $AT^*(t_s, T)$ = \$12.1818, respectively, provided that $\frac{\partial^2 TC}{\partial t_s^2} = 13.3808 > 0$, $\frac{\partial^2 TC}{\partial T^2} = 8.05957 > 0$ and $\frac{\partial^2 TC}{\partial t_s^2} \times \frac{\partial^2 TC}{\partial T^2} - \left(\frac{\partial^2 TC}{\partial t_s \partial T}\right)^2 = 76.6251 > 0$.

EXAMPLE 5.2

Let the parameters be assigned with their proper units: $\alpha = 0.01$, $K = 12$, $\gamma = 0.03$, $D_0 = 8$, $\delta_0 = 2$, $c_1 = 0.6$, $c_2 = 1.5$, $c_o = 10$, $c_s = 2.5$, and $c_l = 2$. Solving Eqs. (5.17) and (5.18), we get $t_s^* = 1.71769$ unit time and $T^* = 2.03625$ unit time. The EOQ and optimal cost function are $I_0^* = 22.1543$ units and $AT^*(t_s, T) =$ \$10.1185, respectively, provided that $\frac{\partial^2 TC}{\partial t_s^2} = 12.3966 > 0$, $\frac{\partial^2 TC}{\partial T^2} = 9.52821 > 0$ and $\frac{\partial^2 TC}{\partial t_s^2} \times \frac{\partial^2 TC}{\partial T^2} - \left(\frac{\partial^2 TC}{\partial t_s \partial T}\right)^2 = 51.056 > 0$.

5.5 SENSITIVITY ANALYSIS

The sensitivity or post optimality analysis on the system parameters are considered by changing +40%, +10%, −10% and −40% of each parameter and are illustrated in the Example 5.1. The notable points are discussed from Table 5.1 as follows:

(i) t_s^* and T^* increase, and $AT^*(t_s, T)$ declines with the decreasing value of α. Here t_s^*, T^* and $AT^*(t_s, T)$ are insensitive sensitive to change in α.
(ii) t_s^* and T^* increase, and $AT^*(t_s, T)$ declines with the decreasing value of K, c_1 and c_2. Here t_s^*, T^* and $AT^*(t_s, T)$ are moderately sensitive to change in K, c_1 and c_2.
(iii) t_s^* and T^* decrease and $AT^*(t_s, T)$ increases with declining value of γ. Here t_s^*, T^* and $AT^*(t_s, T)$ all are high sensitive to change in γ.
(iv) t_s^* and $AT^*(t_s, T)$ decrease, and T^* increases with the decreasing value of D_0, δ_0, c_s, and c_l. Here t_s^*, T^*, and $AT^*(t_s, T)$ all show moderate sensitivity to change in D_0, δ_0, c_s, and c_l.
(v) t_s^*, T^*, and $AT^*(t_s, T)$ all decrease with the decreasing value of c_o. Here all functions show the moderate sensitivity to change in c_o.

5.6 CONCLUDING REMARKS AND FUTURE SCOPES

The traditional EOQ inventory model assumes only the constant demand rate, but not the other factors such as deterioration and shortages. In a real life situation, the demand pattern is affected by the shortages. In this study, an effort has been made to investigate the inventory model for the item that deteriorates according to the time-proportional deterioration rate, exponentially declining demand function, and backlogged shortages. The inspiration of taking on the present model is the thought of both varying deterioration rate and varying demand rate, which can be applied for newly launched items such as fashion items, cosmetics, android mobiles, super computers, laptops, automobiles and their spare parts, and more. The shortages in this system are backlogged and its rate is reciprocal to the length of delay time for the subsequent order. The focus of the study is to optimize the cost function by determining the shortages time point and cycle length of the proposed system as parameters.

TABLE 5.1
Sensitivity Analysis or Post-optimality Analysis

System Parameters	(%) Variation in System Parameters	t_s^*	T^*	$AT^*(t_s, T)$	(%) Variation in $AT^*(t_s, T)$
α	+40	1.28769	1.74236	12.1922	+0.085373
	+10	1.29047	1.74459	12.1844	+0.021343
	−10	1.29234	1.74608	12.1792	−0.021343
	−40	1.29517	1.74835	12.1713	−0.086194
K	+40	1.03380	1.60906	13.6958	+12.4284
	+10	1.21451	1.70017	12.6137	+3.54545
	−10	1.38021	1.80142	11.7052	−3.91239
	−40	1.76268	2.07917	9.92318	−18.5409
γ	+40	1.30267	1.75419	12.1478	−0.279105
	+10	1.29416	1.74749	12.1734	−0.068955
	−10	1.28869	1.74321	12.1901	+0.068134
	−40	1.28078	1.73703	12.2144	+0.267612
D_0	+40	1.34750	1.62199	12.7024	+4.27359
	+10	1.30991	1.70068	12.3536	+1.41030
	−10	1.26799	1.80810	11.9643	−1.78545
	−40	1.18422	2.10181	11.1852	−8.18106
δ_0	+40	1.31108	1.73193	12.3645	+1.49978
	+10	1.29687	1.74188	12.2325	+0.35463
	−10	1.28549	1.74885	12.1269	−0.45067
	−40	1.26435	1.75973	11.9304	−2.06373
c_1	+40	1.03499	1.60986	13.6916	+12.3939
	+10	1.21502	1.70055	12.6121	+3.53232
	−10	1.37943	1.80079	11.7075	−3.89351
	−40	1.75530	2.07264	9.9393	−18.4086
c_2	+40	1.28892	1.74339	12.1894	+0.062388
	+10	1.29078	1.74484	12.1837	+0.015597
	−10	1.29203	1.74582	12.1799	−0.015597
	−40	1.29391	1.74730	12.1742	−0.062388
c_o	+40	1.51409	2.14137	14.2453	+16.9392
	+10	1.35141	1.84664	12.7387	+4.57157
	−10	1.22785	1.64156	11.5912	−4.84822
	−40	1.00914	1.30677	9.55237	−21.5849
c_s	+40	1.31776	1.68295	12.4265	+2.00873
	+10	1.29883	1.72690	12.2508	+0.56641
	−10	1.28329	1.76626	12.1064	−0.61895
	−40	1.25382	1.84978	11.8326	−2.86657
c_l	+40	1.33130	1.65402	12.5522	+3.04060
	+10	1.30345	1.71580	12.2936	+0.91776
	−10	1.27746	1.78184	12.0523	−1.06306
	−40	1.21869	1.96640	11.5059	−5.54844

The effectiveness of the system is discussed with consideration of a couple of numerical examples and sensitivity analysis of deterioration, demand, cost, and cost parameters from one of the two examples is carried out.

There are numerous scopes for further extension for research. For illustration, the model can be generalized by relaxing the restriction of the time-proportional deterioration to several Weibull distributions and Gamma distribution deteriorations. Also, we may extend the model by incorporating the generalized demand functions such as stock-dependent, time-and-price dependent, time-dependent cubic demand rate, and so on. It may also be extended to fuzzy inventory system and stochastic inventory system. Finally, the model can be generalized by incorporating some realistic features like quantity discounts, two warehouses with multi-items system, and production inventory systems.

APPENDIX A

$$\frac{\partial^2 AT(t_s,T)}{\partial t_s^2}=\frac{1}{T}\left[c_1K\left(1-2\gamma t_s+\alpha t_s^2\right)+c_2K\left(-\gamma+\alpha t_s+\gamma e^{-\gamma t_s}\right)+\frac{D_0\left(c_4+\delta_0c_5\right)}{\left(1+\delta_0\left(T-t_s\right)\right)^2}\right].$$

APPENDIX B

$$\frac{\partial^2 AT(t_s,T)}{\partial T^2}=\frac{1}{T}\left[\frac{D_0\left(c_4+\delta_0c_5\right)}{\left(1+\delta_0\left(T-t_s\right)\right)^2}\right]-\frac{2}{T^2}\left[\frac{D_0\left(c_4+\delta_0c_5\right)\left(T-t_s\right)}{1+\delta_0\left(T-t_s\right)}\right]$$
$$+\frac{2K}{T^3}\left[c_1\left(\frac{t_s^2}{2}-\frac{\gamma t_s^3}{3}+\frac{\alpha t_s^4}{12}\right)+c_2\left[t_s-\frac{\gamma t_s^2}{2}+\frac{\alpha t_s^3}{6}+\frac{1}{\gamma}\left(e^{-\gamma t_s}-1\right)\right]\right]+\frac{2c_3}{T^3}$$
$$+\frac{2D_0\left(c_4+\delta_0c_5\right)}{T^3\delta_0}\left[T-t_s-\frac{1}{\delta_0}\ln\left[1+\delta_0\left(T-t_s\right)\right]\right]$$

APPENDIX C

$$\frac{\partial^2 AT(t_s,T)}{\partial t_s\partial T}=-\frac{1}{T}\left[\frac{D_0\left(c_4+\delta_0c_5\right)}{\left(1+\delta_0\left(T-t_s\right)\right)^2}\right]-\frac{K}{T^2}\left[c_1\left(t_s-\gamma t_s^2+\frac{\alpha t_s^3}{3}\right)+c_2\left(1-\gamma t_s+\frac{\alpha t_s^2}{2}-e^{-\gamma t_s}\right)\right]$$
$$+\frac{1}{T^2}\left[\frac{D_0\left(c_4+\delta_0c_5\right)\left(T-t_s\right)}{1+\delta_0\left(T-t_s\right)}\right]$$

REFERENCES

Bag, S., Yadav, G., Dhamija, P., & Kataria, K. K. (2021). Key resources for industry 4.0 adoption and its effect on sustainable production and circular economy: An empirical study. *Journal of Cleaner Production*, *281*, 125233.

Bakker, M., Riezebos, J., & Teunter, R. H. (2012). Review of inventory system with deterioration since 2001. *European Journal of Operational Research*, *221*, 275–284.

Basu, M., Senapati, S., & Banerjee, K. (2006). A multi-item inventory model for deteriorating items under inflation and permissible delay in payments with exponential declining demand. *Opsearch*, *43*(1), 71–87.

Chang, H.J., Dye, C.Y., 1999. An EOQ model for deteriorating items with time varying demand and partial backlogging. *Journal of the Operational Research Society*, *50*, 1176–1182.

Covert, R. P., & Philip, G. C., (1973). An EOQ model for items with Weibull distribution deterioration. *AIIE Transaction*, *5*, 323–326.

Ghare, P. M., & Schrader, G. P. (1963). A model for exponentially decaying inventories. *Journal of Industrial Engineering*, *14*, 238–243.

Ghosh, S. K., & Chaudhuri, K. S. (2006). An EOQ model with a quadratic demand, time proportional deterioration and shortages in all cycles. *International Journal of Systems Sciences*, *37*(10), 663–672.

Hollier, R. H., & Mak, K. L. (1983). Inventory replenishment policies for deteriorating items in a declining market. *International Journal of Production Research*, *21*, 813–826.

Jalan, A., Giri. R., Chaudhury, K. S. (1996). EOQ model for items with Weibull distribution deterioration shortages and trended demand. *International Journal of Production Economics*, *113*, 852–861.

Lakshmi, M. D., & Pandian, P. (2021). Production inventory model with exponential demand rate and exponentially declining deterioration. *Italian Journal of Pure and Applied Mathematics*, *45*, 59–71.

Li, R., Lan, H., & Mawhinney, J. R. (2010). A review on inventory study. *Journal of Service Science and Management*, *3*(1), 117–129.

Mohanty, S., & Singh, T. (2023). A model on an EOQ optimal ordering policy varying with time-dependent cubic demand and variable deterioration under delay in payment conditions. *International Journal of Mathematics in Operational Research*, *26*(1), 37–58.

Mohanty, S., Singh, T., & Routary, S. S. (2022). An optimal replenishment policy with exponential declining demand. *Industry 4.0 and Climate Change*, 169–176.

Mohanty, S., Singh, T., & Sitha, S. (2023). An optimal ordering policy varying with generalized exponential increasing demand, variable deterioration and time-dependent holding cost. *Lecture Notes in Networks and Systems*, *650*, 429–440.

Ouyang, L. Y., Wu, K. S., & Cheng, M. C. (2005). An inventory model for deteriorating items with exponential declining demand and partial backlogging. *Yugoslav Journal of Operations Research*, *15*(2), 277–288.

Raafat, F. (1991). Survey of literature on continuously deteriorating inventory model. *Journal of the Operational Research Society*, *42*, 27–37.

Singh, T., & Pattanayak, H. (2012). An EOQ model for a deteriorating item with time dependent exponentially declining demand under permissible delay in payment. *IOSR Journal of Mathematics*, *2*, 30–37.

Singh, T., Pattanayak, H., Nayak, A. K., & Sethy, N. N. (2018). An optimal policy with three-parameter Weibull distribution deterioration, quadratic demand, and salvage value under partial backlogging. *International Journal of Rough Sets and Data Analysis*, *5*(1), 79–98.

Singh, T., Sethy, N. N., Nayak, A. K., & Pattnaik, H. (2021). An optimal policy for deteriorating items with generalized deterioration, trapezoidal-type demand, and shortages. *International Journal of Information Systems and Supply Chain Management*, *14*(1), 23–54.

Su, C. T., Lin, C. W., & Tsai, C. H., 1999. A deterministic production inventory model for deteriorating items with an exponential declining demand. *Opsearch*, *36*(2), 95–106.

Swain, P., Mallick, C., Singh, T., Mishra, P. J., & Pattanayak, H. (2021). Formulation of an optimal ordering policy with generalised time-dependent demand, quadratic holding cost and partial backlogging. *Journal of Information and Optimization Sciences*, *42*(5), 1163–1179.

Wee, H. M. (1995). A deterministic lot-size inventory model for deteriorating items with shortages and a declining market. *Computer & Operations Research*, *22*(3), 345–356.

6 Development of Sustainability Indicators for Smart and Connected Digital Manufacturing Systems

Patrick Dichabeng, Yaone Rapitsenyane, Richie Moalosi, Victor Ruele, and Oanthata Sealetsa
University of Botswana, Gaborone, Botswana

6.1 INTRODUCTION

The evolution from traditional to digital manufacturing signifies a pivotal transformation in industrial practices, primarily driven by Industry 4.0 or the Fourth Industrial Revolution. This shift began with the automation of individual machines and processes and has since evolved into integrating the entire production systems, altering the manufacturing landscape. The journey towards digital manufacturing can be traced back to the initial stages of automation, where the focus was on enhancing efficiency and reducing human error in isolated tasks. However, the fundamental shift occurred with the development of additive manufacturing (3D Printing), which revolutionised how objects are designed, developed, and manufactured. Additive manufacturing (AM) enabled the creation of complex geometries and customised products with reduced material waste and shorter lead times, marking a significant departure from conventional subtractive manufacturing methods (Mahesh et al., 2021).

Digital manufacturing is exponentially growing in industry to achieve efficiency in producing products and services in terms of cost and environmental footprint (Gregori et al., 2017). The industry now favours the reduced use of physical prototypes or iterations in decision-making about a product for its flexibility in design, analysis, and performance optimisation of factories (Choi et al., 2015; Behera et al., 2023). Digital manufacturing systems (DMS) embody the integration of cutting-edge digital technologies within manufacturing processes, aiming to boost productivity, efficiency, and adaptability in production significantly (Gregori et al., 2017). At the heart of DMS lies the seamless integration of the tangible manufacturing world with the intangible digital realm. This integration is primarily driven by the following technologies:

DOI: 10.1201/9781003349877-6

Artificial intelligence (AI), Internet of Things (IoT), robotics, and big data analytics. These technologies not only help to enhance process control and automation but also enable predictive maintenance, quality control, and supply chain optimisation.

Smart and connected systems are being implemented to improve competitiveness in manufacturing. The introduction of the IoT and Internet of Services in manufacturing has enabled the creation of smart factories with integrated production systems (Gerrikagoitia et al., 2019; Chhotaray et al., 2024). In the IoT framework, physical objects (such as machinery) are linked to the internet, enabling remote access to sensor data and allowing for the distant control of manufacturing equipment, an embedded system (Kopetz and Steiner (2022). This makes the manufacturing machinery a smart object that can receive and process information without human interference. At a basic level, the IoS is meant to facilitate and enable vendors to offer services through the Internet in an end-to-end service environment consisting of providers and consumers, infrastructure, business models, and the services being provided and consumed (Buxmann et al., 2009).

Moreover, integrating cyber-physical systems (CPS) and the broader trend towards Industry 4.0 have necessitated a rethinking of traditional manufacturing paradigms. CPS, which combines the digital and physical elements of manufacturing, has led to the development of more smart and interconnected production assets (Meindl & Mendonça, 2021). This integration is not only about technology but also involves reimagining the roles of humans and machines in a collaborative manufacturing environment, where decision-making is enhanced through AI-powered systems.

Regarding sustainability, energy flows, costs, and data management or product/process information management are vital factors to be monitored in smart and connected digital manufacturing systems to minimise emissions and hazardous materials during product development (Kurniadi & Ryu, 2021). A systematic literature review by Andronie et al. (2021) shows that sustainable smart and connected digital manufacturing integrates complete input of product life cycle procedures. The life cycle perspective allows for identifying potential sustainability indicators from extracting raw materials to applying end-of-life strategies. There is also a need to consider social sustainability in manufacturing to indicate how smart manufacturing affects work organisations regarding social interaction and organisational structure (Cagliano et al., 2019). This chapter looks at smart and connected digital manufacturing systems. It proposes sustainability indicators across the product life cycle to enable real-time decision-making regarding environmental impacts, economic viability, and socio-cognitive implications.

6.2 SUSTAINABILITY INDICATORS IN MANUFACTURING SYSTEMS

Sustainability assessment is increasingly recognised as an important instrument to facilitate the transition towards sustainable practices (Gładysz & Kluczek, 2019). It is an essential part of sustainable development. Moldavskaa and Welob (2015) argue that if the level of sustainability cannot be measured at an organisation level, we do not know if we are doing the right things or achieving sustainability goals. Sustainability assessment helps to achieve the triple-bottom line goals of the environmental, economic, and social factors (Chand, 2020). Some assessment studies concentrate on environmental and economic impacts and fail to reflect on social considerations or simultaneously integrate the three goals (Gladysz et al., 2020; Moldavskaa & Welob, 2015). All the social, economic, and environmental considerations were taken on board in this chapter. Sustainability indicators in digital manufacturing systems are metrics used to assess and monitor digital manufacturing processes and operations' environmental, social, and economic impacts (Moldavskaa & Welob, 2015). These indicators help organisations identify areas for improvement and make informed decisions to achieve a more sustainable and responsible manufacturing system.

6.2.1 Economic Sustainability Indicators

In digital manufacturing systems, economic sustainability indicators evaluate and quantify manufacturing activities' profitability, effectiveness, and financial well-being in digitalisation. Digital manufacturing uses cutting-edge technologies to streamline operations, cut expenses, boost output, and allocate resources as efficiently as possible. Table 6.1 shows a synthesis of economic sustainability indicators in digital manufacturing systems developed by the following scholars: Javaid et al. (2022), Contini and Peruzzini (2022), Sherif et al. (2022), Embia et al. (2023), Chand (2020), Gładysz & Kluczek (2019), and Hristov and Chirico (2019).

In the context of digital manufacturing systems, economic sustainability metrics assist firms in maximising their financial performance and achieving long-term economic sustainability. Ongoing observation and evaluation of these indicators facilitate well-informed decision-making and stimulate enhancements to augment operational effectiveness and financial gain (Table 6.1).

6.2.2 Environmental Sustainability Indicators

Environmental sustainability indicators in digital manufacturing systems are critical for assessing and mitigating the environmental impact of manufacturing processes enabled by digital technology. Digital manufacturing aims to minimise environmental effects by maximising resource use, cutting waste, and improving eco-efficiency. Table 6.2 illustrates the environmental indicators that should be considered in digital manufacturing systems as synthesised from Ferreira et al. (2023), Javaid et al. (2022), Contini and Peruzzini (2022), Sherif et al. (2022), Chand (2020), Gładysz and Kluczek (2019), and Hristov and Chirico (2019).

Monitoring and optimising specific environmental sustainability metrics can reduce detrimental environmental effects and promote responsible resource use in digital manufacturing systems. These measures are necessary to balance environmental stewardship, economic development, and human welfare.

6.2.3 Social Sustainability Indicators

Social sustainability indicators in digital manufacturing systems assess the impact on stakeholders, employees, and society. In digital manufacturing, fair labour practices, community involvement, and employee well-being should be prioritised as per the

TABLE 6.1
Economic Sustainability Indicators in Digital Manufacturing Systems

Criteria	Indicator	Description
Economic	Production cost	Track cost of producing products and services, direct and indirect costs. Identify areas for cost reduction and efficiency improvement.
	Operational efficiency	Evaluate the efficiency of manufacturing operations, optimise resource utilisation, and minimise waste.
	Resource utilisation and optimisation	Monitor and optimise the usage of resources to achieve higher productivity and cost efficiency whilst minimising waste. Increase revenue associated with the sustainability dimension.
	Return on investment	Assess and invest in digital technologies vs. anticipated financial benefits.
	Process automation and labour productivity	Measure labour productivity gains achieved through process automation and digital technologies vs. manual.
	Manufacturing flexibility and agility	Assess digital manufacturing systems' ability to adapt to changing market demands, product variations, and production schedules.
	Production scalability	Ensure optimal utilisation of resources. Evaluate the ease and cost of scaling production up or down based on market demand and business needs.
	Energy efficiency and cost	Track energy consumption and costs associated with digital manufacturing processes. Identify opportunities to enhance energy efficiency and reduce expenses.
	Sustainability cost savings	Quantify cost savings from sustainability initiatives, e.g., reduced waste, lower resource consumption, and improved eco-efficiency.
	Innovation and technology adoption	Measure the level of innovation and technology adoption. Ensure that digital manufacturing systems remain up to date and competitive.
	Customer value and market competitiveness	Assess the value delivered to customers through enhanced product quality, customisation, and shorter lead times.

indicators in Table 6.3 (Ferreira et al., 2023; Piso et al., 2023; Javaid et al., 2022; Contini and Peruzzini, 2022); Sherif et al., 2022; Chand, 2020; Gładysz & Kluczek, 2019; Hristov & Chirico, 2019).

Monitoring and improving social sustainability indicators in digital manufacturing systems will promote a more ethical, inclusive, and socially responsible approach to manufacturing, which will benefit the local workforce and communities. Economic, environmental, and social sustainability indicators assist manufacturing systems in meeting sustainability development goals, lessening environmental impact, becoming more socially responsible, and helping maintain an entity's long-term economic viability. These indicators support a sustainable and resilient industry when integrated into the production process.

TABLE 6.2
Environmental Sustainability Indicators in Digital Manufacturing Systems

Criteria	Indicator	Description
Environmental	Carbon emissions and footprint	Measure and track greenhouse gas emissions generated by manufacturing operations.
	Energy efficiency and renewable energy use	Monitor energy consumption in digital manufacturing processes. Evaluate the integration and utilisation of renewable energy sources to enhance energy efficiency and reduce the use of fossil fuels.
	Water usage and conservation	Track water consumption in manufacturing processes. Implement measures to optimise water usage, reduce waste, and promote water conservation.
	Material efficiency and waste reduction	Assess the efficient use of raw materials and track waste generation. Minimise waste through improved production processes and circular economy practices. Reduce the use of natural resources.
	Life-cycle assessment (LCA)	Conduct a comprehensive LCA of products manufactured using digital technologies. Analyse the environmental impact across the entire product life cycle (from raw material extraction to disposal).
	Toxic substance and hazardous material use	Monitor and reduce the use of toxic substances, hazardous materials, and gases in manufacturing processes. Minimise damage to the environment and human health.
	Circular economy integration	Evaluate the adoption of circular economy principles within digital manufacturing systems.
	Eco-friendly supply chain	Assess suppliers' environmental sustainability practices to promote eco-friendly manufacturing and transportation practices.
	Waste management and recycling	Monitor and manage the disposal of waste generated during the use and end-of-life of digital manufacturing equipment and components.
	Air quality and emissions control	Monitor air emissions from digital manufacturing operations. Implement measures to control and reduce air pollutants, enhancing air quality in the environment.
	Sustainable packaging	Evaluate and optimise packaging materials and designs to minimise environmental impact. Reduce packaging waste and improve upcycling and recyclability.
	Software optimisation for efficiency	Optimise software algorithms and models to reduce computational resource requirements, thus minimising energy consumption.
	Smart manufacturing for optimisation	Leverage smart manufacturing technologies to optimise production schedules, equipment utilisation, and energy consumption, thus reducing environmental impact.
	Biodiversity and ecosystem impact	Assess and mitigate the impact of manufacturing operations on local biodiversity, ecosystems, and natural habitats to preserve and protect environmental diversity.

TABLE 6.3
Social Sustainability Indicators in Digital Manufacturing Systems

Criteria	Indicator	Description
Social	Employee health and safety	Monitor and ensure compliance with occupational health and safety regulations to minimise workplace accidents, injuries, and occupational illnesses.
	Employee well-being and work-life balance	Evaluate employee satisfaction, mental health, work hours, and work-life balance to promote a healthy and positive work environment.
	Diversity and inclusion	Track workforce diversity in gender, ethnicity, age, and other factors to ensure an inclusive and equitable workplace that values and respects diversity.
	Training and skill development	Implement strategies to offer training and development programs for employees, enhancing their skills and knowledge, which in turn promotes career advancement and increases job satisfaction.
	Fair wages and benefits	Assess whether employees receive fair wages and benefits that satisfy or exceed legal and industry standards to promote financial well-being and stability.
	Labour rights and unionisation	Monitor compliance with labour rights and freedom of association, including the right to unionise and collective bargaining, to ensure fair treatment of workers.
	Community engagement and impact	Evaluate engagement with local communities, including support for community programmes, employment prospects for residents, and minimising adverse impacts on the community.
	Ethical sourcing and supply chain responsibility	Assess the ethical sourcing of materials and components, ensuring they are produced under fair labour conditions and adhere to social responsibility standards.
	Customer and consumer satisfaction	Measure customer satisfaction and feedback to ensure that products and services meet consumer expectations and needs.
	Stakeholder engagement and collaboration	Engage with various stakeholders (e.g., NGOs, local organisations, government) to ensure that the manufacturing system considers their concerns and needs, promoting collaboration and understanding.
	Conflict-free sourcing	Ensure that raw materials and components are sourced in a way that does not contribute to conflict, human rights abuses, or unethical practices.
	Knowledge sharing and collaboration	Encourage knowledge sharing and collaboration among employees, industry peers, and academia to foster innovation, knowledge transfer, and collective growth.
	Employee involvement in decision-making	Involve employees in decision-making processes that impact their work, providing opportunities for input and participation.
	Responsible automation and workforce transition	Consider the effect of automation on the workforce and implement strategies to retrain and upskill employees, ensuring a smooth transition to new roles or technologies.

6.3 SMART AND CONNECTED DIGITAL MANUFACTURING SYSTEMS

Smart manufacturing, or Industry 4.0, has garnered substantial attention from researchers, engineers, and policymakers (Oztemel & Gursev, 2020; Qu et al., 2019; Zheng et al., 2018; Galgal et al., 2023; Abubakr et al., 2020). Scholars and experts have contributed various definitions of smart manufacturing systems, reflecting their diverse and evolving nature (Abubakr et al., 2020; Qu et al., 2019; Zheng et al., 2018). For example, an enhanced application of cutting-edge intelligence systems that enable the quick manufacture of new items, dynamic reaction to product demand, real-time supply chain network, and manufacturing production optimisation is known as smart manufacturing (Qu et al., 2019). A common theme across this definition is the integration of cutting-edge technologies into manufacturing processes to enhance efficiency, flexibility, and productivity. Smart refers to the system's ability to gather and analyse data in real time, make informed decisions, and communicate with other systems and components within the manufacturing ecosystem. As a result, it transforms conventional manufacturing into an interconnected, data-driven, and intelligent domain.

The foundational elements of digital manufacturing pivot around three critical parts: cyber-physical systems (CPS), digital twins, and smart factories. CPS are integrations that blend computational (digital) and physical processes. In a manufacturing context, this means machines and systems equipped with sensors and interconnected through networks, enabling real-time data collection and analysis. This data-driven approach enables more responsive and adaptable manufacturing processes, where physical changes in the system can be immediately reflected and managed digitally. Digital twins take this concept further by creating high-fidelity virtual models of physical systems. These models serve as real-time digital counterparts of physical entities, providing a platform for simulation, analysis, and control. Digital twins can predict the performance, maintenance needs, and operational efficiencies of their physical counterparts, thereby enabling pre-emptive actions and decision-making to optimise the manufacturing process. They are particularly valuable in complex systems where physical testing is impractical or too costly (Park et al., 2019).

Smart factories represent the culmination of these technologies into a cohesive, digitised, and connected production facility. In a smart factory, all components, from supply chains to production lines, are interconnected and intelligent. This interconnectivity ensures that the manufacturing process is not only automated but also adaptable to changing conditions and demands. Smart factories leverage IoT for real-time monitoring and control, AI for decision-making and process optimisation, and robotics for automation and precision. The integration of these technologies results in a manufacturing environment that is not only efficient and productive but also resilient and sustainable. The IoT further propelled this evolution. IoT's integration into manufacturing systems enabled the seamless collection and exchange of data across various components of the production line, leading to more informed decision-making and predictive maintenance strategies. This connectivity not only enhanced operational efficiency but also enabled the development of more agile and responsive manufacturing ecosystems.

The advent of machine learning and AI in manufacturing has been another cornerstone of this transformation. AI algorithms and machine learning techniques are now employed to optimise production processes, improve supply chain management, and enhance product quality. These technologies have enabled manufacturers to predict and pre-emptively address potential issues, thereby reducing downtime and enhancing productivity (Masum, 2023). This rapid digitalisation and interconnectivity have also introduced new challenges, particularly cybersecurity. With manufacturing systems increasingly reliant on digital technologies, they have become more vulnerable to cyber threats (Roy et al., 2023). These threats range from stealing sensitive information to sabotaging products or production lines, leading to financial and reputational damages. Therefore, securing digital manufacturing environments against such threats has become a critical aspect of modern manufacturing strategies.

6.4 SUSTAINABILITY AND SMART AND CONNECTED DIGITAL MANUFACTURING SYSTEMS

There is an urgent need to transition from a linear economy based on linear production chains, end-of-life waste, wasteful energy usage, ecosystem degradation, resource losses, and wasteful utilisation of products (Wu & Pi, 2023). The circular economy promises to address some of the challenges experienced by humanity and the environment. In a circular economy (CE), the typical business model is the product-service system (PSS), where users pay for a product's functioning (Wu & Pi, 2023). This change in business tactics marks a move from a product-oriented approach aimed at maximizing sales to a service-oriented strategy designed to generate profits through the provision of services (Kjaer et al., 2019). Digital technology and manufacturing systems are regarded as critical enablers of a PSS and make it easier for enterprises to integrate CE into their sustainable business models.

Sustainability and smart, networked digital manufacturing systems go hand in hand because they support more productive, ecologically conscious, and financially sustainable manufacturing methods. The following are some critical ways that the application of smart and networked digital manufacturing systems promotes sustainability. Pirola et al. (2020) advance that sustainability and smart, networked digital manufacturing systems result in resource efficiency. That is, maximising the use of resources by using real-time data and monitoring. This involves minimising water use, reducing material waste, and using energy efficiently. To achieve sustainability, digital manufacturing systems must produce more with less. Turner et al. (2021) propose that sustainable digital manufacturing systems lead to a diminished environmental footprint. Digital manufacturing systems can lessen their environmental impact, including pollutants and greenhouse gas emissions, by optimising production schedules and procedures based on real-time data. Such systems are aligned with the circular economy principles. Recycling, remanufacturing, and waste reduction are just a few of the circular economy's tenets that smart manufacturing supports. Valuable materials can be recovered from end-of-life products using digital systems that track and manage product lifecycles (Wu & Pi, 2023).

Sustainable digital manufacturing systems can be designed to support energy management. Renewable energy sources can be integrated into connected production systems, enabling precise control over energy consumption (Tsunetomo et al., 2022). This lowers the manufacturing process' carbon impact.

Using sustainable manufacturing techniques can result in lower costs, more productivity, and enhanced competitiveness (Pirola et al., 2020), thus leading to the economic viability of the manufacturing entity. The sustainability of manufacturing operations is thereby guaranteed. On another dimension, digital technologies make it possible to perform thorough life-cycle assessments and analyses of goods and procedures, which can be used to pinpoint areas that need to be improved for sustainability.

Smart and networked digital manufacturing systems greatly aid the advancement of sustainability in the manufacturing sector. Ultimately, these methods promote resource efficiency, lessen environmental effects, and increase economic sustainability, which results in a more conscientious and sustainable method of producing goods. However, digital manufacturing systems must take social and ethical values into account (Tsunetomo et al., 2022).

The evolution of smart and connected manufacturing systems shows that the current trend of using Industry 4.0 technologies in manufacturing uses sustainability and customer participation as integral components of its business model (Qu et al., 2019). The sustainability objective requires a holistic view of managing and controlling not only the value creation networks but also the entire product life cycle and the product development process from design to manufacturing (Oertwig et al., 2017; Brown et al., 2014). This helps to achieve the resource efficiency sustainability objective in self-organising manufacturing systems.

6.5 CONCEPTUAL MODEL OF SMART MANUFACTURING

The transformative potential of Industry 4.0 and the Fourth Industrial Revolution (4IR) technologies in advancing sustainable development within the manufacturing sector is captured in the layered approach diagram (See Figure 6.1) as a three-tiered model. It illustrates the foundational role of digital technologies, the subsequent process improvements they facilitate, and the overarching sustainable outcomes they foster. Figure 6.1 serves as a conceptual map, guiding the reader through the interplay between smart manufacturing technologies and their contributions to economic, environmental, and social sustainability.

The following description delves into each layer of the model, elucidating the dynamic ways in which digitalisation can amplify sustainable development performance in the manufacturing realm.

- **Base Layer (Technology)**: The core is up of foundational 4IR technologies such as AI, the IoT, CPS, and so on. These technologies are the bedrock of smart manufacturing, enabling the collection and analysis of data, automating complex tasks, and fostering a more interconnected and intelligent manufacturing environment.
- **Middle Layer (Process Improvements)**: The second layer showcases the enhancements in manufacturing processes attributable to 4IR technologies. These include gains in efficiency through more precise control and reduction of waste; flexibility with systems that can adapt to changes in demand

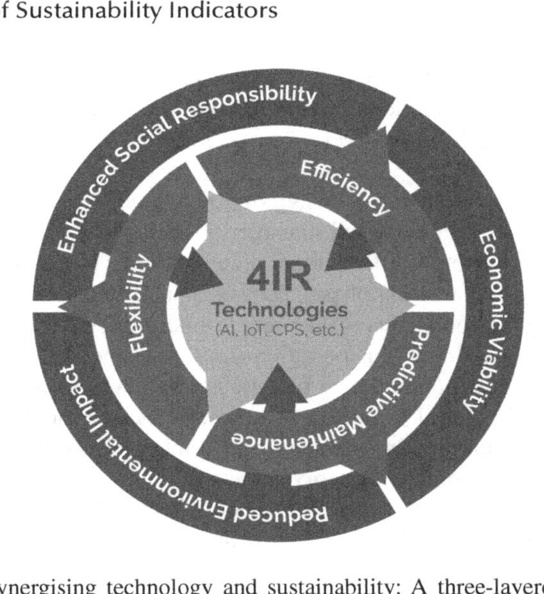

FIGURE 6.1 Synergising technology and sustainability: A three-layered model of smart manufacturing.

or material supply; and predictive maintenance, allowing for anticipatory upkeep to prevent downtime.

- **Outer Layer (Sustainability Outcomes)**: The outer layer represents the sustainability outcomes that arise from the integration of digital technology in manufacturing processes. These outcomes include a reduced environmental impact due to more efficient resource use and waste reduction; enhanced social responsibility, as intelligent systems can improve worker safety and create higher quality jobs; and economic viability, with cost savings from optimised processes and new opportunities for growth.

The arrows feeding back into the central core signify the iterative nature of improvement: data from improved processes and sustainability outcomes inform further technological advancements, creating a continuous loop of enhancement. Figure 6.1 encapsulates how digital technologies not only drive improvements in manufacturing processes but also significantly boost sustainable development performance. The continuous feedback loop ensures ongoing progress, with the potential for digital technologies to provide even more significant improvements as they evolve and become more integrated into manufacturing ecosystems.

6.6 IMPLICATIONS FOR PRACTICE

AM plays a crucial role in digital manufacturing systems (DMS), facilitating the swift and economical creation of intricate and bespoke designs. This significantly cuts down on material waste and allows for production tailored to immediate demand. This flexibility in manufacturing is crucial for industries that require high customisation or produce in small batches, offering a competitive edge in the market (Kumar et al., 2023). However, the transition to DMS has its challenges. One of the primary hurdles is the high cost of implementation. The upfront cost of investing in advanced technologies like AI, IoT, and robotics can be significant, especially for small and medium-sized businesses.

Additionally, the complexity of integrating new technologies with existing systems poses significant technical and managerial challenges. This complexity often requires a strategic overhaul of traditional manufacturing processes and can lead to disruptions during the transition phase. The sustainability benefits of AM are, however, enormous. Once companies make parts in-house, they can reduce their carbon footprint as less material and transportation of parts are needed. In AM, there is no material-wasting process (subtractive manufacturing), which means parts are built with optimal use of material. Complex sub-assemblies and components can also be produced as one part, eliminating the need for assembly procedures and related costs. This further saves the company's energy and costs.

Another significant challenge is data security. With the increasing reliance on interconnected systems and the IoT, digital manufacturing has become a prime target for cyber threats. These threats include everything from solitary hackers to organized criminal groups and state-sponsored actors, posing threats to the secrecy and integrity of sensitive manufacturing data (Masum, 2023). The cybersecurity risks extend to the entire digital manufacturing ecosystem, including supply chains and production lines, where malicious attacks can cause financial and reputational damage (Roy et al., 2023). While these challenges exist, digital sustainability comes along with Industry 4.0, which requires that companies manage and utilise digital resources in a way that makes them available in the long term and supports their secure accessibility. Responsible data practices will compel companies to monitor data usability, validity, duplication, and necessity.

Workforce training is another critical challenge. The shift to digital manufacturing requires a workforce skilled in new technologies and processes. This need for specialised skills necessitates significant investment in training and development, which can be a barrier, especially for smaller manufacturers. Furthermore, the development and operationalisation of these systems involve intricate challenges, particularly in accurately capturing and modelling the dynamics of physical phenomena (Malik, 2021). The creation of virtual models and simulations, integral to digital twins, requires a deep understanding of the manufacturing process and the ability to adapt these models throughout the life cycle of the factory. To operate smart factories, companies must develop 21st-century soft skills simultaneously with Industry 4.0 skills to benefit the workforce within the smart manufacturing environments.

6.7 CONCLUSION

Sustainability in manufacturing encapsulates a multifaceted approach, encompassing the judicious management of resources, responsible environmental stewardship, and the cultivation of ethical labour practices (Herrmann et al., 2014). A vital component of the concept put forth by the World Commission on Environment and Development (1987) is the commitment to meet current needs without sacrificing the opportunity of future generations to meet their needs. At its core, sustainability in manufacturing acknowledges that businesses play a critical role in shaping the world's ecological and social well-being. The shift toward smart and connected digital manufacturing signifies a transformative step in amplifying sustainability across environmental, social, and economic domains. To actualise this shift, embedding comprehensive sustainability indicators within these modern production systems becomes imperative. Economic metrics such as cost, resource utilisation, energy efficiency, and investment returns should be dynamically analysed for financial optimisation. Environmental measures, including carbon emissions, material use efficiency, renewable energy integration, and waste reduction, are crucial for curbing ecological impacts. Social factors, such as worker health and safety, skills development, inclusivity, and community involvement, must be consistently monitored to maintain ethical and people-centric practices.

Utilising advanced analytics, AI, and digital twins, these sustainability indicators can be systematically tracked and forecasted, enabling a proactive and data-informed approach to refine the triple bottom line—people, planet, and profit. However, the journey to sustainable smart manufacturing is not a solitary one. It demands collective action and policy transformation to institute universal sustainability reporting standards. The embedding of these principles must start from the design phase, extending through the entire product life cycle and resonating with the circular economy values. In essence, placing sustainability at the forefront of smart digital manufacturing systems is pivotal. Such integration not only balances economic progress with environmental care and social welfare but also charts a course for an ethically sound, responsible, and sustainable manufacturing future. The successful adoption of these technologies and principles globally will not only reshape manufacturing processes but also have the potential to significantly impact the sustainability of our planet and societies, echoing the true spirit and goals of Industry 4.0.

The integration of sustainability principles should commence from the design phase, traverse the entire product lifecycle, and resonate with the ethos of the circular economy. In essence, positioning sustainability at the core of smart digital manufacturing systems not only reconciles economic advancement with environmental stewardship and social welfare but also paves the way for an ethically responsible and sustainable future. The global embrace of these technologies and principles has the potential to reshape manufacturing processes and, more significantly, impact the sustainability of our planet and societies. It embodies the genuine spirit and aspirations of Industry 4.0, offering a beacon of hope for a brighter, more sustainable tomorrow.

REFERENCES

Abubakr, M., Abbas, A. T., Tomaz, Í., Soliman, M. S., Luqman, M., & Hegab, H. (2020). Sustainable and smart manufacturing: An integrated approach. *Sustainability*, 12(6), 2280. https://doi.org/10.3390/su12062280

Andronie M, Lăzăroiu G, Ștefănescu R, Uță C, & Dijmărescu I. (2021). Sustainable, smart, and sensing technologies for cyber-physical manufacturing systems: A systematic literature review. *Sustainability*, 13(10), 5495. https://doi.org/10.3390/su13105495

Behera, B.C., Moharana, B.R., Rout, M. and Debnath, K., (2023). Application of machine learning in the machining processes: Future perspective towards Industry 4.0. *Intelligent Manufacturing Management Systems: Operational Applications of Evolutionary Digital Technologies in Mechanical and Industrial Engineering*, pp. 141–156.

Brown, A., Amundson, J., & Badurdeen, F. (2014). Sustainable value stream mapping (Sus-VSM) in different manufacturing system configurations: application case studies. *Journal of Cleaner Production*, pp. 85, 164–179.

Buxmann, P., Hess, T., & Ruggaber, R. (2009). Internet of services. *Business & Information Systems Engineering*, 1, 341–342.

Cagliano, R., Canterino, F., Longoni, A., & Bartezzaghi, E. (2019). The interplay between smart manufacturing technologies and work organisation: The role of technological complexity. *International Journal of Operations & Production Management*, 39(6/7/8), 913–934.

Chand, A. (2020). Sustainability indicators in the dairy industry. *Nature Food*, pp. 1, 397. https://doi.org/10.1038/s43016-020-0122-x

Choi, S., Jun, C., Zhao, W. B., & Do Noh, S. (2015). Digital manufacturing in smart manufacturing systems: contribution, barriers, and future directions. In *Advances in Production Management Systems: Innovative Production Management Towards Sustainable Growth: IFIP WG 5.7 International Conference*, APMS 2015, Tokyo, Japan, September 7–9, 2015, Proceedings, Part II 0 (pp. 21–29). Springer International Publishing.

Chhotaray, P., Behera, B.C., Moharana, B.R., Muduli, K., & Sephyrin, F.T.R., (2024). Enhancement of Manufacturing Sector Performance with the Application of Industrial Internet of Things (IIoT). In *Smart Technologies for Improved Performance of Manufacturing Systems and Services*, pp. 1–19. CRC Press.

Contini, G. & Peruzzini, M. (2022). Sustainability and Industry 4.0: Definition of a Set of Key Performance Indicators for Manufacturing Companies. *Sustainability*, 14, 11004. https://doi.org/10.3390/su141711004

Embia, G., Mohamed, A., Moharana, B. R., & Muduli, K. (2023). Edge Computing-Based Conditional Monitoring. *Intelligent Manufacturing Management Systems: Operational Applications of Evolutionary Digital Technologies in Mechanical and Industrial Engineering*, 249–270.

Ferreira, J. J., Lopes, J. M., Gomes, S. & Rammal, H. G. (2023). Industry 4.0 implementation: Environmental and social sustainability in manufacturing multinational enterprises. *Journal of Cleaner Production*, 404, 136841. https://doi.org/10.1016/j.jclepro.2023.136841

Galgal, K.N., Ray, M., Moharana, B.R., Behera, B.C. and Muduli, K., (2023). Quality Control in the Era of IoT and Automation in the Context of Developing Nations. In *Smart Technologies for Improved Performance of Manufacturing Systems and Services*, pp. 39–50. CRC Press.

Gerrikagoitia, J. K., Unamuno, G., Urkia, E., & Serna, A. (2019). Digital manufacturing platforms in the industry 4.0 from private and public perspectives. *Applied Sciences*, 9(14), 2934.

Gładysz, B., Kluczek, A. (2019). An indicators framework for sustainability assessment of RFID systems in manufacturing. In: Trojanowska, J., Ciszak, O., Machado, J., Pavlenko, I. (eds) *Advances in Manufacturing II. MANUFACTURING 2019. Lecture Notes in Mechanical Engineering*. Springer, Cham. https://doi.org/10.1007/978-3-030-18715-6_23

Gladysz, B., Ejsmont, K., Kluczek, A., Corti, D., & Marciniak, S. (2020). A method for an integrated sustainability assessment of RFID technology. *Resources*, 9, 107. https://doi.org/10.3390/resources9090107

Gregori, F., Papetti, A., Pandolfi, M., Peruzzini, M., & Germani, M. (2017). Digital manufacturing systems: A framework to improve social sustainability of a production site. *Procedia CIRP*, 63, 436–442. https://doi.org/10.1016/j.procir.2017.03.113

Herrmann, C., Schmidt, C., Kurle, D., Blume, S., & Thiede, S. (2014). Sustainability in manufacturing and factories of the future. *International Journal of Precision Engineering and Manufacturing-Green Technology*, 1, 283–292.

Hristov, I., & Chirico, A. (2019). The role of sustainability Key Performance Indicators (KPIs) in implementing sustainable strategies. *Sustainability*, 11, 5742. https://doi.org/10.3390/su11205742; www.mdpi.com/journal/sustainability

Javaid, M., Haleem, A., Singh, R. P., Khan, S. & Suman, R. (2022). Sustainability 4.0 and its applications in the field of manufacturing. *Internet of Things and Cyber-Physical Systems*, 2, 82–90. https://doi.org/10.1016/j.iotcps.2022.06.001

Kjaer, L. L., Pigosso, D. C., Niero, M., Bech, N. M., & McAloone, T. C. (2019). Product/service systems for a circular economy: The route to decoupling economic growth from resource consumption? *Journal of Industrial Ecology*, 23(1), 22–35.

Kopetz, H., & Steiner, W. (2022). Internet of Things. In *Real-time systems: design principles for distributed embedded applications* (pp. 325–341). Cham: Springer International Publishing.

Kumar, M., Epiphaniou, G., & Maple, C. (2023). Leveraging Semantic Relationships to Prioritise Indicators of Compromise in Additive Manufacturing Systems. Paper presented at the *Applied Cryptography and Network Security Workshops*, Cham.

Kurniadi, K. A., & Ryu, K. (2021). Development of multi-disciplinary green-BOM to maintain sustainability in reconfigurable manufacturing systems. *Sustainability*, 13(17), 9533.

Mahesh, P., Tiwari, A., Jin, C., Kumar, P. R., Reddy, A. L. N., Bukkapatanam, S. T. S., … Karri, R. (2021). A survey of cybersecurity of digital manufacturing. *Proceedings of the IEEE*, 109(4), 495–516. https://doi.org/10.1109/JPROC.2020.3032074

Malik, A. A. (2021). *Framework to model virtual factories: a digital twin view*. abs/2104.03034. https://arxiv.org/ftp/arxiv/papers/2104/2104.03034.pdf

Masum, R. (2023). *Cyber Security in Smart Manufacturing (Threats, Landscapes Challenges)*. abs/2304.10180. https://arxiv.org/pdf/2304.10180.pdf

Meindl, B., & Mendonça, J. (2021). *Mapping Industry 4.0 Technologies: From Cyber-Physical Systems to Artificial Intelligence*. abs/2111.14168.

Moldavskaa, A. & Welob, T. (2015). On the applicability of sustainability assessment tools in manufacturing. *Procedia CIRP*, 29, 621–626. https://doi.org/10.1016/j.procir.2015.02.203

Oertwig, N., Jochem, R., & Knothe, T. (2017). Sustainability in model-based planning and control of global value creation networks. *Procedia Manufacturing*, 8, 183–190.

Oztemel, E. & Gursev, S. (2020). Literature review of Industry 4.0 and related technologies. *Journal of Intelligent Manufacturing*, 3(1), 127–182. https://doi.org/10.1007/s10845-018-1433-8

Park, H., Easwaran, A., & Andalam, S. (2019). Challenges in Digital Twin Development for Cyber-Physical Production Systems. Paper presented at the *Cyber Physical Systems. Model-Based Design*, Cham.

Pirola, F., Boucher, X., Wiesner, S. & Pezzotta, G. (2020). Digital technologies in product-service systems: A literature review and a research agenda, *Computers in Industry*, 123. https://doi.org/10.1016/j.compind.2020.103301

Piso, K., Mohamed, A., Moharana, B. R., Muduli, K., & Muhammad, N. (2023). Sustainable Manufacturing Practices through Additive Manufacturing: A Case Study on a Can-Making Manufacturer. *Intelligent Manufacturing Management Systems: Operational Applications of Evolutionary Digital Technologies in Mechanical and Industrial Engineering*, 349–375.

Qu, Y. J., Ming, X. G., Liu, Z. W., Zhang, X. Y. & Hou, Z. T. (2019). Smart manufacturing systems: state of the art and future trends. *The International Journal of Advanced Manufacturing Technology*, 103(9–12), 3751–3768. https://doi.org/10.1007/s00170-019-03754-7

Roy, P., Bhargava, M., Chang, C.Y., Hui, E., Gupta, N., Karri, R., & Pearce, H. (2023). *A survey of Digital Manufacturing Hardware and Software Trojans*. https://arxiv.org/pdf/2301.10336.pdf

Sherif, Z., Sarfraz, S., Jolly, M., Salonitis, K. (2022). Identification of the right environmental KPIs for manufacturing operations: Towards a continuous sustainability framework. *Materials*, 15, 7690. https://doi.org/10.3390/ma15217690

Tsunetomo, K. Watanabe, K. & Kishita, Y. (2022). Smart product-service systems design process for socially conscious digitalisation, *Journal of Cleaner Production*, 368, 133172. https://doi.org/10.1016/j.jclepro.2022.133172

Turner, C.J., Oyekan, J.O., & Stergioulas, L.K. (2021). Distributed manufacturing: A new digital framework for sustainable modular construction. *Sustainability*, 13(3), 1515. https://doi.org/10.3390/su13031515

WCED (1987). World commission on environment and development. *Our Common Future*, 17(1), 1–91.

Wu, D. & Pi, Y. (2023). Digital technologies and product-service systems: A synergistic approach for manufacturing firms under a circular economy. *Journal of Digital Economy*, 2, 37–49. https://doi.org/10.1016/j.jdec.2023.04.001

Zheng, P., Wang, H., Sang, Z., Zhong, R., Liu, Y., Liu, C., Mubarok, K., Yu, S., & Xu, X. (2018). Smart manufacturing systems for Industry 4.0: Conceptual framework, scenarios, and future perspectives. *Frontiers of Mechanical Engineering*, 13(2), 137–150. https://doi.org/10.1007/s11465-018-0499-5

7 Performance Investigations on Robots to Attain Environmental Sustainability

Shiv Manjaree Gopaliya, Pushpdant Jain, and Suchetana Sadhukhan
VIT Bhopal University, Bhopal, India

Manoj Kumar Gopaliya
The NorthCap University, Gurugram, India

David Chua Sing Ngie
Universiti Malaysia Sarawak, Kota Samarahan, Malaysia

7.1 INTRODUCTION

Environmental sustainability is a major point of concern in today's times. It gives insights and concepts about human-environment interaction in a responsible manner. Many factors such as pollution, deforestation, over-exploitation of natural resources, and even our unsustainable lifestyle leading to health issues have forced us to look into the amount of harm we have created to our environment (Ogbemhe et al. 2017). On the other hand, development in the field of modern robotics contributed to the better handling of issues such as manpower, cost of production, fatigue during work hours, and sustainability as well (Bugmann et al. 2011). Moreover, the sustainability gains from utilizing robots in various sectors can help to achieve enhancement of activities and increase safety, and programmability by semi-skilled workers is easily possible (Project STAMINA 2013; Santana et al. 2007). Researchers are also developing various means such as software and hardware to enhance the sustainable characteristics of different robotics systems (Pellicciari et al. 2015). In conjunction with this, tools such as sustainability framework for smart and associated systems can be utilized (Pan et al. 2018; Rinaldi et al. 2023). These indicators and frameworks provide a systematic approach to assessing and improving the environmental performance of smart and connected systems. The Robobench platform developed by Weisz et al. (2016) aims to systematize and ease the work selection and execution process.

DOI: 10.1201/9781003349877-7

Similarly, Stroupe et al. (2006) have created a multi-robot system to minimize the use of various resources in the field of construction, showcasing that sustainable practices can be incorporated into robotic systems. The authors concluded that despite various uncertainties and sensing and motions this system performs several tasks very easily. When it comes to other applications robots are also utilized in the restaurant business utility by considering the various aspects related to customer safety, intelligence, animacy, and likability of the system (Jang and Soo-Bum 2020). Researchers are also redesigning various robots based on the type of industries, where some robots are designed for larger-scale operations while others are optimized for small or medium-scale industries (Bi et al. 2015). Author Dias et al. (2005) have employed the details of various robots in the field of education in countries like Ghana, the USA, and Sri Lanka to utilize their contribution to the higher education system. Industrial applications like servicing robots are also performing various tasks (Kohl et al. 2020). New requirements of safety sensors for service robots working near humans were proposed (Schraft and Merklinger 1996). Other field applications include the harsh environment robot's application as proposed by Bausys et al. (1996); the authors concluded that a multi-criterion approach is better for dynamic decision-making. Shehu and Nuhu (2019) have applied robotics application in the field of civil engineering development of building applications in terms of energy efficiency, reduced maintenance cost, enhanced air quality, and so forth. An analysis was carried out in the form of a case study in which an electric fan was redesigned and a new robot-based assembly line was proposed (Vast et al. 1997). One of the major reasons for the imbalance in the environment is industrial development. With Industry 4.0 in place, robots have gained a lot of importance in manufacturing industries. Various applications handled by industrial robots can be seen in Figure 7.1. Robots are the most advanced technological machines (Wang et al. 2018). Robots are also assumed to be unsustainable machines. The role of

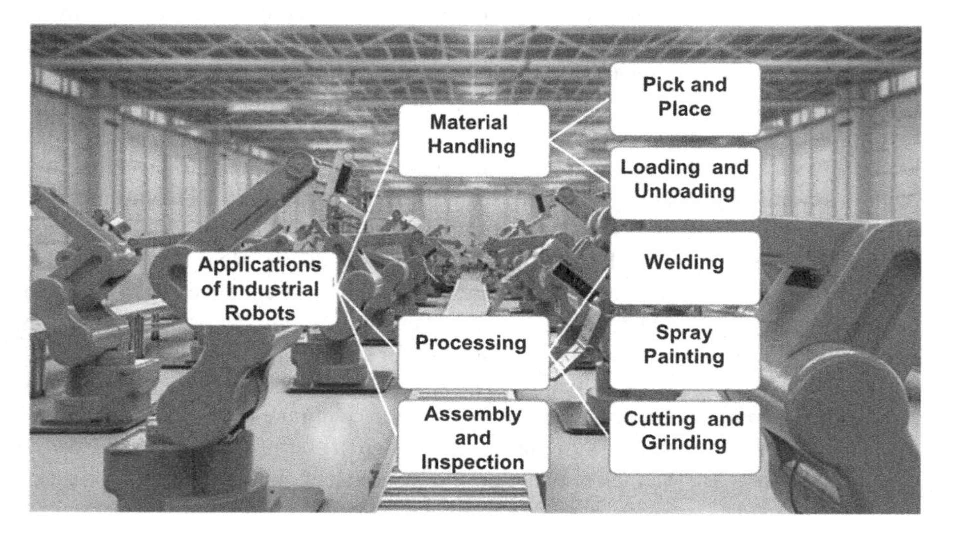

FIGURE 7.1 Applications of industrial robots.

robots in helping towards environmental sustainability is a question to be answered by the new researchers. A comprehensive investigation has been made by Chen et al. (2022) on industrial robots and their ecological footprint highlighting the dual impact, both on the economy and the environment across 72 countries. However, present studies are highly counterproductive and need more attention. Investigations carried out by Gong et al. (2023) explored the impact of robots on the economy benefiting the aging population. They highlight that more attention needs to be paid to the economic influence of robots and data gaps within aging societies. The energy efficiency of industrial robots leading to cost savings and performance improvements contributes towards sustainable practices in part manufacturing (Soori et al. 2022). The study highlights that focusing on sustainable practices in part manufacturing in the productivity enhancement domain also presents promising opportunities for future research.

How can the COVID-19 pandemic prompt a shift in focus towards lifestyle choices and sustainability? It underscores the significance of preserving energy and the environment. While robots are typically perceived as unsustainable, this endeavor seeks to explore the unaddressed query regarding their capacity to promote sustainability by means of inventive designs, novel concepts, innovative materials, and advanced technologies. Can the adoption of a sustainability framework serve as a roadmap for creating eco-friendly robots? This chapter is an attempt to highlight that robots can also be treated as part of an environmentally sustainable solution. An attempt is also made here to provide different ways in which a robot can be treated as a sustainable solution.

7.2 ROBOTS: UNSUSTAINABLE MACHINE

Industrial growth has seen the maximum use of the best technologically advanced machines. Robots have the major advantages of fine quality and quantity. Irrespective of all the advantages, robots are believed to be unsustainable machines due to two reasons: (a) amount of energy consumption and (b) unemployment issues. The reasons have been discussed in detail in the following sections.

7.2.1 Robots Consume Lots of Energy

There is an exceptional gap in the rate of energy consumption between a developed country and a developing country. Robots work on electricity. The amount of energy consumed is a point of concern for robots. There is also a pre-existing misconception that robots consume a lot of energy while performing any task. To move towards a full sustainability solution, the consumption of energy by industrial robots is one of the major criteria to be handled with expertise. One of the ways to reduce energy consumption by robots is by selecting the proper pattern of end-effector movement for any given application. By changing the acceleration and velocity we can save the energy of a robot end effector subjected to selection of the appropriate robot for a particular application. It will be better to go for complete computer simulations of industrial applications before working on the actual robotic manipulator. By applying a combination of given methods, energy can be saved leading to sustainability.

7.2.2 Robots Are the Cause of Unemployment

One of the major points of argument is always the rise in unemployment due to the deployment of robots in industries. As said earlier, robots are the most advanced form of machines resulting in products of high quality and also in good quantity. A broad range of skill sets is needed for finding employment in robot-based industries. Lack of employment and lack of skill set are two separate issues to be addressed. Based on the latter, it is the need of the hour to educate and prepare our present and future generations with multiple industry-specific skill sets. It can be achieved when both industries and academics work together with the target to bridge the skill gap. While working on robots, skilled manpower in small numbers can also serve the purpose. In other words, lots of scopes are available to study and find new jobs for unemployed manpower so that a balance in environmental sustainability can be achieved.

Integrating sustainability indicators and frameworks guides the development and deployment of robotics systems. Indicators quantify energy efficiency, resource consumption, emissions, and waste management, allowing analysis of environmental impacts and identification of improvement areas. Dedicated tools facilitate sustainable practices, including energy consumption simulation, optimization algorithms for resource-efficient path planning, and data analysis platforms for tracking environmental performance. By incorporating sustainability criteria, such as optimizing end-effector movement patterns and utilizing simulations, robots can reduce energy consumption and promote sustainability. Informed decisions by researchers, engineers, and manufacturers further enhance the environmental sustainability of robots and their applications. This chapter discusses and highlights the importance of indicators in the light of environmental sustainability in the following sections.

7.3 CASE STUDY: IMPORTANCE OF PROPER END-EFFECTOR MOVEMENT OF THE ROBOT

The industrial revolution has seen the maximum use of robots. The end-effector of a robot is the most vital part of any industrial application. While selecting the right type of robot is important for a particular industrial application, it is also of utmost importance to check the proper movement of the end-effector. Industrial applications like pick and place or welding or palletizing heavily depend on the proper end effector motion, which in turn brings it to the required position. It involves both kinematics and dynamics analyses for the robot under study.

Kinematic analysis of a robot is a very crucial step to understanding the end-effector motion properly. Kinematic analysis can be accomplished in different ways such as forward and inverse kinematics. The first helps to identify the coordinates of the end-effector of a robot at any given point in time using the Denavit-Hartenberg convention. Manseur (1996) proposed a software program named Kinematics Analysis Program for five- or six-axis robots of general geometry. Koyuncu and Guzel (2007) have provided theoretical analyses of the 5-degree-of-freedom (DOF) robot and to test the motion properties they have also developed a software package named MSG. Several textbooks in robotics (Niku 2001; Saha 2008; Mittal and Nagrath 2003) provide forward and inverse kinematic and dynamic analyses of multi-DOF robots.

The joint angles at any given time for a particular coordinate can be identified through inverse kinematics. The availability of multiple solutions makes inverse kinematics a complex task as compared to forward kinematics. The inverse kinematic solution can be achieved by either a simulation approach, geometrical approach, or experimental approach. Yahya et al. (2011) have suggested a geometrical technique using inverse kinematics to optimize the path plan of the hyperredundant manipulator's end effector. Sheng et al. (2005) have proposed a geometrical approach to obtain inverse kinematic solutions for a serially connected n-link hyperredundant planar robotic manipulator. Bayraktaroglu et al. (1999) have proposed a method for the movement of a snake-like mechanism and simulation results have been given for the trajectory followed and the body form. Pravak's manual (2008) provides information regarding the physical parameters of the 3-DOF robot under study.

To highlight the concept of environmental sustainability as the need of the hour, the importance of proper end-effector movement of the robot has been described experimentally using a 3-DOF industrial pick-and-place robot. Figure 7.2 shows a 5-DOF pick-and-place type robot with a gripper as its end-effector. It has 3 degrees of freedom in the links and 2 degrees of freedom in the spherical wrist. However, it has been used as a 3-DOF planar robot by restricting the movements at the wrist. A spherical wrist allows the decomposition of the 5-DOF robot into 3 degrees of freedom and 2 degrees of freedom sub-problems, respectively. The particular movement of each joint of the robot is described in Table 7.1.

Where, J and W represent joint and wrist respectively. L, R, F, B, U, D, ST, Et, C, and AC are considered as left, right, forward, backward, up, down, sky turn, earth turn, clockwise, and anticlockwise respectively. The actual specifications of the three links of the used 3-DOF robot under study are link length $L_0 = 226$ mm, link length $L_1 = 179$ mm, link length $L_2 = 177$ mm with joint angles as $-90° \leq \theta_0 \leq 90°$, $0° \leq \theta_1 \leq 90°$, $0° \leq \theta_2 \leq 90°$.

FIGURE 7.2 3-DOF robot with gripper.

TABLE 7.1
Motion Characteristics of the 3 Degrees of Freedom Robot

Type of Joint	Part of Robot	Type of Movement
J - 1	Waist	L/R
J - 2	Shoulder	F/B
J - 3	Elbow	U/D
W	Wrist pitch	ST/ET
W	Wrist roll	C/AC

The forward kinematic equations for the investigated 3-DOF robot are provided as follows:

$$x = l_0 C_1 + l_1 C_{12} + l_2 C_{123} \tag{7.1}$$

$$y = l_0 S_1 + l_1 S_{12} + l_2 S_{123} \tag{7.2}$$

$$\phi = \theta_1 + \theta_2 + \theta_3 \tag{7.3}$$

The inverse kinematic equations for the 3-DOF robot under study are as follows:

$$\theta_1 = a\tan 2\left(\left(k_1 y_n - k_2 x_n\right), \left(k_1 x_n - k_2 y_n\right)\right) \tag{7.4}$$

$$\theta_2 = a\tan 2\left(S_2, C_2\right) \tag{7.5}$$

$$\theta_3 = \phi - \left(\theta_1 + \theta_2\right) \tag{7.6}$$

where, $k_1 = l_0 + l_1 C_2$, $k_2 = l_1 S_2$, $C_2 = \dfrac{x^2 + y^2 - l_0^2 - l_1^2}{2 l_0 l_1}$, $S_2 = \pm\sqrt{\left(1 - C_2^2\right)^2}$, $x_n = x - l_2 C\phi$, $y_n = y - l_2 S\phi$

Torques and forces exerted on individual joints of the industrial robot are responsible for the movement of the robot for a particular application. The equations of motion concerned with the relationship between the forces and accelerations acting on a robot at any given time are called robot dynamics. If the robot's physical parameters are known, a dynamic study of the robot can be performed. Robot dynamics can be done in two ways, which are forward dynamics and inverse dynamics. Forward dynamics aids in comprehending how robots react to given forces or torques. Inverse dynamics assists in determining the necessary actuator forces and torques needed to produce the desired trajectory of the robot's end-effector. As compared to forward dynamics, inverse dynamics is of utmost importance because a proper inverse dynamic mathematical model helps in the real-time control of robot movement. Conkur (2003) has introduced an algorithm for path planning in highly redundant manipulators, employing approximated B-spline curves. The path following is defined by ensuring that the manipulator links remain approximately tangent to these curves. Arakelian et al. (2011) have conducted an analytically tractable solution for minimizing input torques in a 2-degree-of-freedom serial manipulator, employing minimum energy control and optimal redistribution of movable masses. The proposed approach was validated through numerical simulations conducted using ADAMS software. Cheah and Liaw (2004) have formulated and resolved two inverse Jacobian regulators with gravity compensations, addressing stability issues. The theoretical findings were experimentally verified by implementing the inverse Jacobian regulators on an industrial robot, specifically the PUMA 560.

Inverse dynamics analysis gives us a better understanding of the joint angle, velocity, and acceleration of each link of the robot. For any robot-based industrial application such as different types of welding, assembling, placing the object at various places, and so on, the proper movement and positioning of the end-effector of the robot can be managed by manipulating the velocity and acceleration of each robot joint. In other words, minimizing error in end-effector positioning will in turn help in minimizing the energy consumption by the robot, leading to sustainable solutions.

Considering the various parameters of the manipulator, computer simulations have been carried out and inverse dynamic analysis has been performed for the 3-DOF robot under study. Figures 7.3a to 7.5a show the plots for joint angle, velocity, and acceleration, and Figures 7.3b to 7.5b plot the torque for each joint of the 3-DOF manipulators. It can be seen from the plots that the torque required to move link 1 is quite different as compared to other links of the 3-DOF robot. Based on the physical structure of the 3-DOF robot, the speed variation of link 1 is a little different. The speed variations of link 2 and link 3 are similar, resulting in the proper end-effector movement on the desired trajectory.

In this study, a desired 'straight line' trajectory has been plotted experimentally using 3-DOF robot with a gripper. 'Straight line' trajectory is the simplest trajectory that can be plotted using a multi-DOF robot where the initial and final coordinates of the end-effector of the robot are known. Table 7.2 gives the coordinates of 'straight line' trajectory. Figure 7.6 shows the actual 'straight line' trajectory plotted experimentally with the help of 3-DOF robot.

Here, the joint angles have been obtained using the 'stepper motor motion' of each joint, which is given as follows:

$$\text{Joint angle } \theta_1 \text{ for link 1 motion}$$
$$= \frac{\left(\text{angle moved by one step movement, i.e., 7.5 degree}\right) \times \text{(difference of distance moved from initial position)}}{\text{(diameter of gear responsible for each movement of robotic manipulator)}}$$

Based on the physical parameters of the 3-DOF robot, the length of the trajectory has been calculated using MATLAB simulations. Similarly, the experimentally plotted length of trajectory has been measured. In conclusion a comparison is performed between the analytical approach and experimental approach for plotted trajectory. The error can be seen in Table 7.3. A maximum of 9% error is obtained while plotting

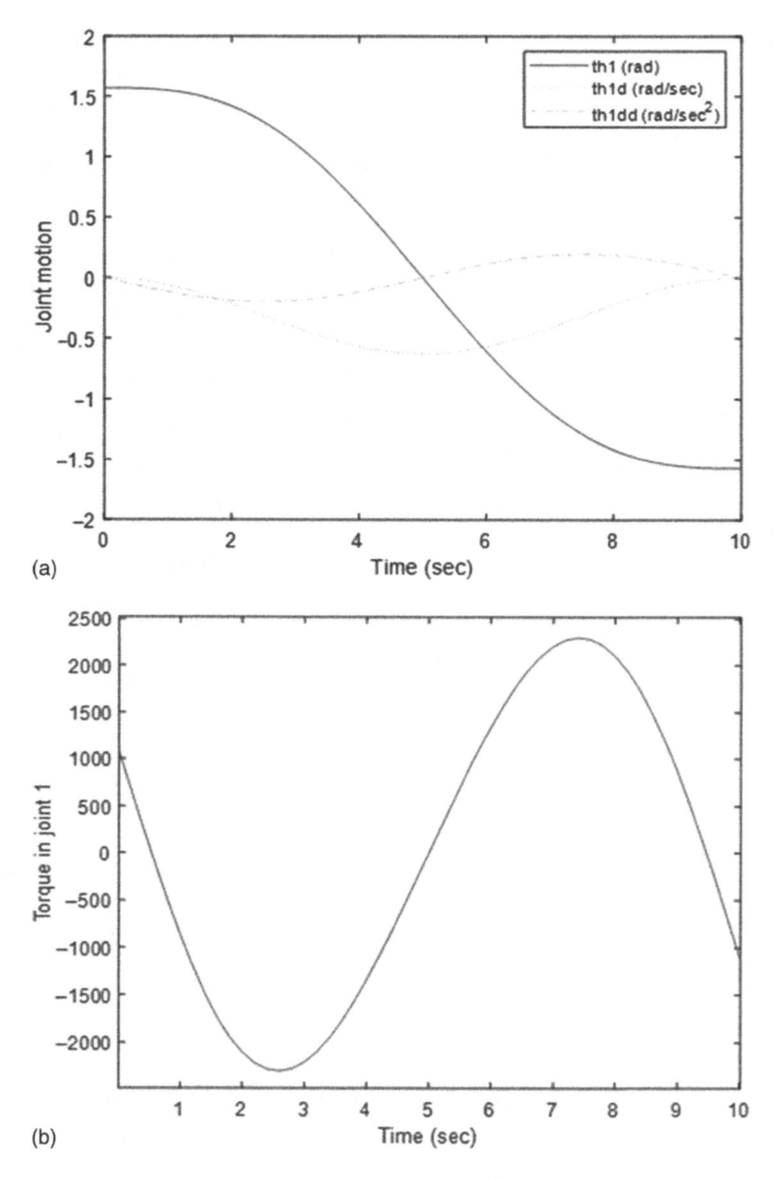

FIGURE 7.3 (a) Joint trajectory for link 1 of 3-DOF robot; and (b) Torque of link 1 of 3-DOF robot.

the trajectory experimentally. Non-rigidity and non-stiffness of robot joints are the major reason for the occurrence of error. The other reasons could be slackness of the gripper, minor vibrations in the gripper while performing the task, or misalignment.

In this case study, forward and inverse kinematics analyses of the 3-DOF robot have been performed. Since environmental sustainability is the concept of the future,

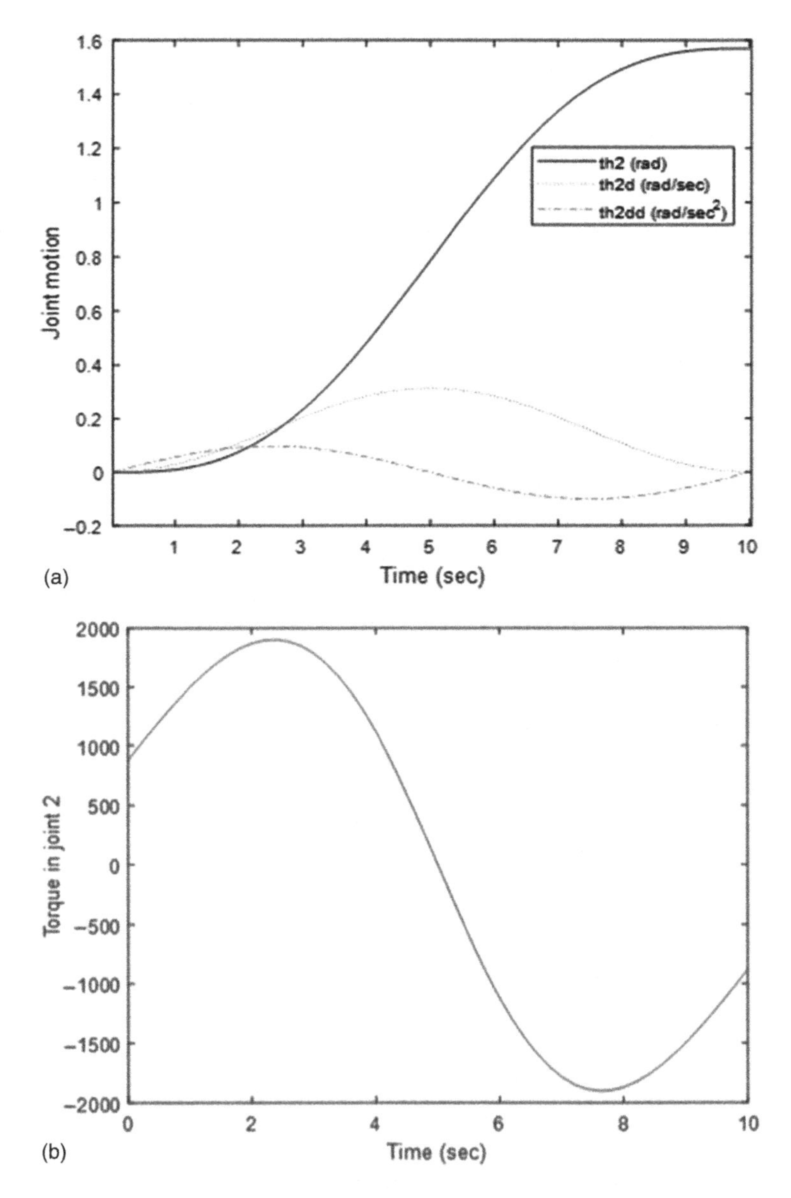

FIGURE 7.4 (a) Joint trajectory for link 2 of 3-DOF robot; and (b) Torque of link 2 of 3-DOF robot.

inverse dynamics analysis has also been performed which helps in a better understanding of velocity, acceleration, and torque for each joint of the robot. The real-time control of dynamic parameters like velocity, acceleration, torque, and force can help in the proper movement of the end-effector of the robot, that is, minimizing energy consumption leading to sustainable solutions.

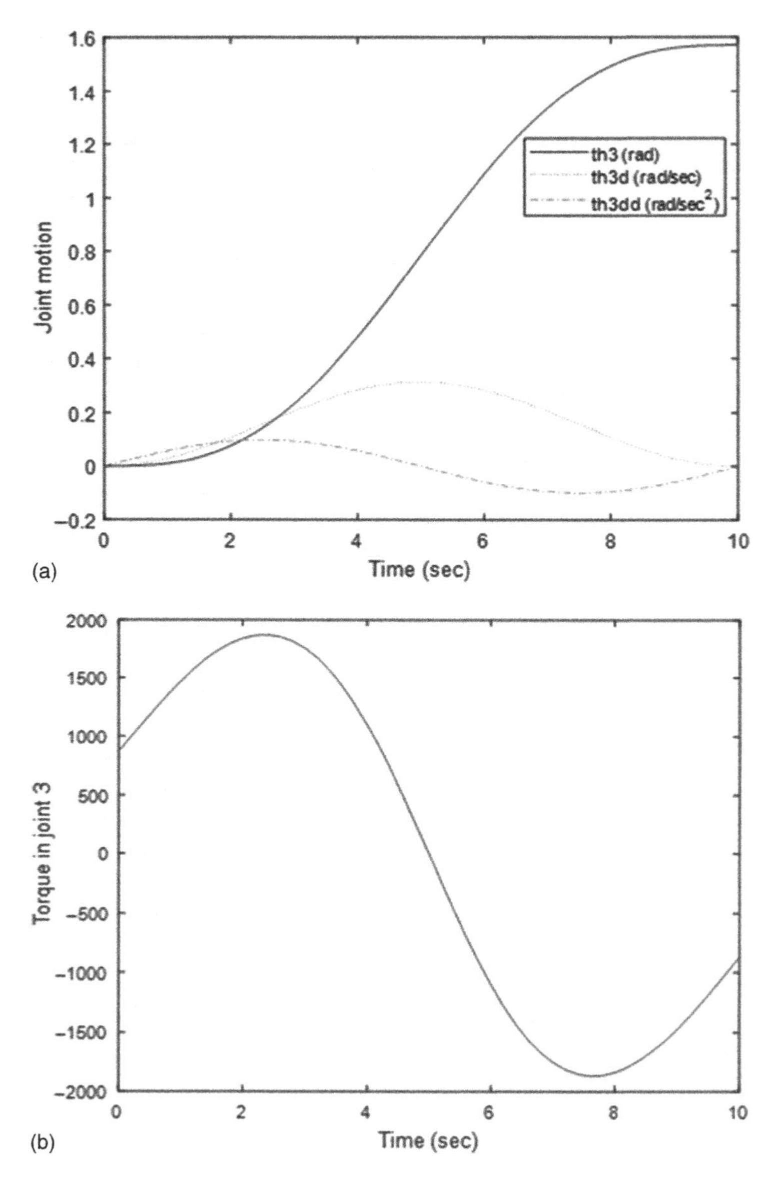

FIGURE 7.5 (a) Joint trajectory for link 3 of 3-DOF robot; and (b) Torque of link 3 of 3-DOF robot.

7.4 NEW APPROACHES TO ROBOTS LEADING TOWARD SUSTAINABILITY

The current study highlights that robots can also be treated as a solution to achieve environmental sustainability. With all the industrial advancements going on in the present scenario, it is the right time to accept that robots can also play an important

TABLE 7.2
Coordinates of 'straight line' Trajectory

Point	X (mm)	Y (mm)
A	403	103
B	349	226
C	305	301
D	290	335

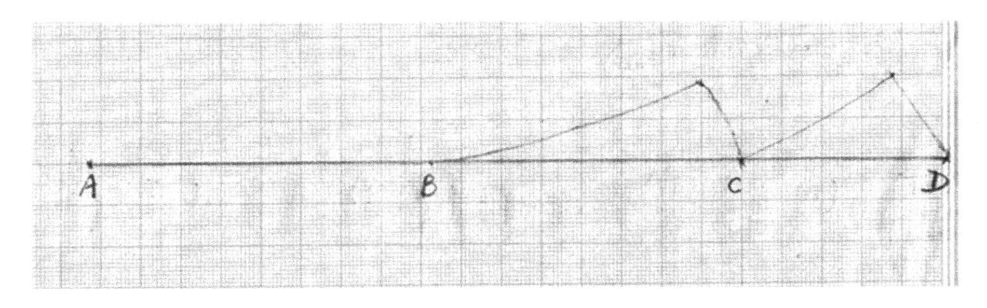

FIGURE 7.6 'Straight line' trajectory plotted using the 3-DOF robot.

TABLE 7.3
Error Obtained for 'straight line' Trajectory

S. No.	Straight Line Segments	Plotted Length (mm) (Experimental)	Forward Kinematics (mm) (MATLAB)	Error (%)
1	AB	85	82	3.5
2	BC	78	71	8.9
3	CD	51	47	7.0

role in the goal of sustainability. As the first step in this study, the robots will be termed sustainable robots. Sustainability has opened ways for many concepts, ideas, and changes to be brought to macro as well as micro levels of the environment. The same approach is needed for sustainable robots too. New approaches to achieving sustainability with the help of robots or manipulators, as depicted in Figure 7.7, are discussed in detail.

7.4.1 New Designs of Robots

The major components of industrial robots are controllers, sensors, robotic manipulators, end effectors, and drive. As sustainability is the point of concern, the robotic components should also be redesigned from their traditional ways. For example, controllers and drives can be redesigned in terms of energy efficiency.

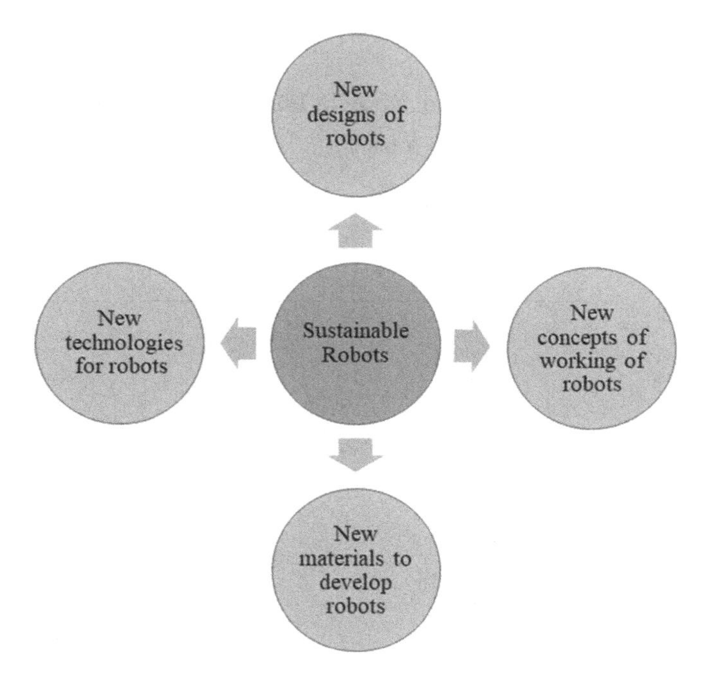

FIGURE 7.7 Various approaches to designing sustainable robots.

Electric motors can be replaced with variable speed motors which help to align the speed of machines with the speed of the assembly line, hence, saving energy. It is seen that maximum energy is consumed in moving the manipulator from one point to another based on the application. New energy-efficient robots can be redesigned as per their optimized movements needed for a particular application. In simpler cases, self-actuated robots can also serve the purpose of saving energy effectively. Sustainability concerning energy consumption is also possible with the help of robots in the ways that robots do not need proper lighting in the workstations and also robots do not need air-conditioning or a heater as per the climatic conditions.

7.4.2 New Concepts of Working with Robots

Unemployment issues can be handled with this approach. The ergonomic working of the man-machine system can be applied in the case of robots too. In other words, the best fit interaction between man and robot can be deduced. For example, in the first step, segregation of tasks performed by a robot can be done. Then the best talents of skilled manpower can be utilized to perform certain tasks while the robot remains idle. This approach may be treated as a win-win solution for environmental sustainability.

7.4.3 New Materials to Develop Robots

Sustainable robots can be designed and developed using sustainable materials. For example, most of the car manufacturing industries are using recycled materials for new models. The same approach should be applied to develop robots. It will help in reducing the use of material. Robots have found their applications in almost every area, such as industries, medical, defense, space and sea explorations, home, and entertainment. Instead of using typical materials like steel or aluminum, it's time to explore and use new materials for the development of robots based on their particular application. For example, alloys or smart materials can be used to make micro or nanorobots needed for invasive surgeries. Robotic end-effectors or grippers can be developed using alloys making them lightweight and flexible.

7.4.4 New Technologies for Robots

With the COVID-19 pandemic creating worldwide havoc, robots have played a crucial role to help humankind in the best way possible. A few years back, when the application of robots was mostly limited to industries, this pandemic situation opened multiple ways for robots to prove their worth. Robots can be seen monitoring patients in hospitals, delivering medicines and door-to-door supplies, cleaning, and helping frontline medical workers. These new robotic applications have also opened doors to search for new techniques or new technologies to perform tasks more efficiently and effectively. Even new robotic application domains can be searched for. It is going to be a very creative and challenging task for new researchers in the near future.

7.5 ROLE OF EDUCATION FOR SUSTAINABLE ROBOTS

Society has become aware of the importance of preserving the environment for our future generations. A lot of steps are being taken in this direction, for example, courses like environmental sustainability have been made compulsory by educational institutions. As said earlier, the pandemic situation has opened multiple ways for robot applications. From high-end industries to small start-ups, many are already thinking of new ways to enhance the usage of robots in day-to-day life. Education plays an important role in increasing the acceptance of robots in human society. It is the right time to teach the design, analysis, and development of sustainable robots. Educational institutions can prepare curricula, syllabi, society-based projects, and training programs to achieve the objective, as represented in Figure 7.8. These steps will help to bridge the gap between the theory and practical aspects of life.

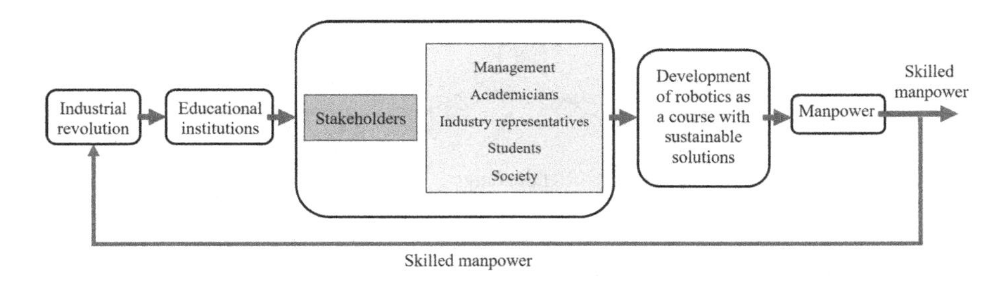

FIGURE 7.8 Education playing an important role in the industrial revolution.

7.6 SUSTAINABLE FRAMEWORK, INDUSTRIES AND ROBOTS

Sustainable framework has three major components; the environment, social, and economy aspects, as shown in Figure 7.9. While industrial revolution has played a crucial part in the development of a country, it is also important to sustain the environment from further harm. Robots are the new machines. The Fourth Industrial Revolution has seen the rise in the use of robots for various applications as shown in Figure 7.1. When compared to humans, robots have proved their significance in many ways. One of the major reasons is precision with speed in its performance. As discussed in Section 7.4, new concepts, designs, technologies, and materials leading to low energy consumption for sustainable robots will help us in preserving the environment.

Robots are best to do repetitive, boring jobs. In other words, the scope of human error is almost negligible while working with robots. Issues like sick leave or striking for a pay rise are not applicable in the case of robots. Robots are not affected by environmental changes such as humidity, temperature, pressure, and so on. However, Industry 5.0 looks into the aspect of human-robot interaction addressing the issues of non-employment. Also, the education sector is playing an important role in disseminating knowledge about the latest technology and machines, that is, robots and artificial intelligence, hence the social criteria of sustainability framework.

Artificial Intelligence is an extra tool that has enhanced the performance adaptability and robustness of robots. It has helped create robots for production with the best quality and good quantity. Robots have the power of decision-making just like humans. This feature is possible only by the interference of artificial intelligence into the machines. Artificial Intelligence enhances the robot's performance by making it more flexible and accurate along with decision-making capabilities. In other words, incorporating computational intelligence will further enhance the sustainability framework performance. All the factors mentioned here make robots the most preferred choice to be a part of the industrial revolution.

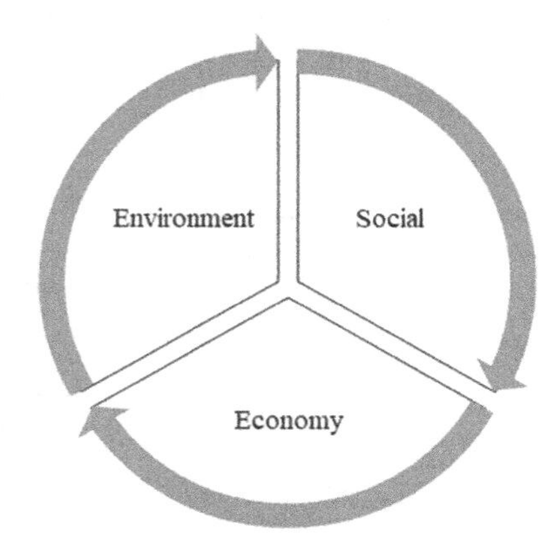

FIGURE 7.9 Sustainable framework.

7.7 FUTURE OF ROBOTS POST-COVID PANDEMIC

While the COVID pandemic is still a threat to humankind, robots are coming up as strong help. Robots are machined away from the risks of any infections. As compared to previous years, this pandemic has opened a new market for robots and robotic industries. Robots are finding their widespread applications in nursing patients in hospitals, cleaning, delivering items, and helping frontline medical workers. Robots are gaining human trust with time. Nowadays, robots are playing a vital role in the medical industry. Robots in different shapes and sizes (viz. micro-bots, nanobots, cobots, etc.) have been developed to serve various medical requirements. Recent times have witnessed doctors performing complicated invasive surgeries with the help of robots. One of the major reasons behind the rise in robot usage is the lack of skilled manpower to complete complicated tasks.

It has also opened new domains/fields of applications for robots. With all the technological advancements in present times, homes are also going 'smart'. In other words, a home where all the devices and machines are convenient to control automatically. For example, smart security systems, smart doorbells, smart lighting systems, and so forth. In a smart home, all the devices are connected and controlled at a central point. The user can control home appliances via a smartphone or laptop. Robots are also playing a major role in making homes 'smart'. Robot vacuum cleaners, robot lawn mowers, and smart irrigation systems are such examples. Robots have moved out of the industrial periphery and have reached our homes. A lot of potential is being seen in robotic home systems as well as robotic entertainment systems. Artificial Intelligence and the latest technologies provide robots with the required performance and flexibility.

There are also different areas of applications such as defense, space, and underwater exploration, entertainment, and agriculture which have seen an increase in the use of robots. All these new applications lead toward sustainable solutions for robot development. The different approaches mentioned in this paper can be used to develop sustainable robots for future applications.

7.8 CONCLUSIONS

Environmental sustainability is the need of the hour. Robots can also play a major role in sustaining the environment. Multiple ways have been provided to handle the issue of energy consumption by industrial robots. In this chapter, it was discussed that robots can also be called sustainable robots. Sustainable robots can be developed by new designs, new concepts, and new technologies as well as by using new materials to fabricate robot components. A case study on desired trajectory planning on a pick-and-place robot has been presented, showcasing the importance of proper movement and positioning of the end-effector of the manipulators. Error in results is mostly due to non-rigidity and non-stiffness of robot joints, loose gripper, misalignment, and minor vibrations in gripper, highlighting the importance of new lightweight, robust designs of the robot to be developed and used which will lead us to sustainable solutions. The results can also be improved by applying artificial intelligence techniques and simulations before working on the real robot. To achieve this goal, education has also a very important role to play. The lack of skill sets can be positively bridged with the help of robotics-based courses, training centers, and society-based projects. In other words, it's time to prepare our generations to study robots, which will help in developing new smart robots for future applications. New areas of applications of robots during the COVID pandemic have been discussed in detail. It is also highlighted to develop sustainable robots. In these times, achieving sustainability with the best available technology, that is, robots, should be the approach to move forward.

REFERENCES

Arakelian, V, J P L Baron, and P Mottu. "Torque Minimization of the 2-DOF Serial Manipulators Based on Minimum Consideration and Optimum Mass Redistribution." *Mechatronics, Elsevier 21*, no. 1 (2011): 310–314. https://doi.org/10.1016/J.MECHATRONICS.2010.11.009

Bausys, R, C Fausto, and S Rokas. "Application of Sustainability Principles for Harsh Environment Exploration by Autonomous Robot." *Sustainability 11*, no. 9 (1996): 2518. https://doi.org/10.3390/SU11092518

Bayraktaroglu, Z Y, F Butel, P Blazevic, and V Pasqui. "A Geometrical Approach to the Trajectory Planning of a Snake like Mechanism." *In IEEE/RSJ International Conference on Intelligent Robots and Systems (IROS'99) 3*, (1999): 1322–1327. https://doi.org/10.1109/IROS.1999.811663

Bi, Z M, Y Liu, B Baumgartner, E Culver, J N Sorokin, A Peters, B Cox, J Hunnicutt, J Yurek, and S O'Shaughnessey. "Reusing Industrial Robots to Achieve Sustainability in Small and Medium-sized Enterprises (SMEs)." *Industrial Robot 42*, no. 3 (2015): 264–273. https://doi.org/10.1108/IR-12-2014-0441

Bugmann, G, S Mel, and B Rachel. "A Role for Robotics in Sustainable Development?" *IEEE Africon'11*, (2011): 1–4. https://doi.org/10.1109/AFRCON.2011.6072154

Cheah, C C, and H C Liaw. "Inverse Jacobian Regulator with Gravity Compensation: Stability and Experiment." *Proceedings of IEEE International Conference on Control Applications 1*, (2004): 321–326. https://doi.org/10.1109/TRO.2005.844674

Chen, Yang, Cheng Liang, and Lee Chien-Chiang. "How Does the Use of Industrial Robots Affect the Ecological Footprint? International Evidence." *Ecological Economics 198*, (2022). https://doi.org/10.1016/j.ecolecon.2022.107483

Conkur, E. S., "Path following Algorithm for Highly Redundant Manipulators, Robotics and Autonomous Systems." *Elsevier 45*, (2003): 1–22. https://doi.org/10.1016/S0921-8890(03)00083-6

Dias, M B, G Ayorkar, and N Thrishantha. "Robotics, Education, and Sustainable Development." *Proceedings of the IEEE International Conference on Robotics and Automation*, (2005): 4248–4253. https://doi.org/10.1109/ROBOT.2005.1570773

Gong, Chi, Yang Xianghui, Tan Hongru, and Lu Xiaoye. "Industrial Robots, Economic Growth, and Sustainable Development in an Aging Society." *Sustainability 15*, no. 5 (2023): 4590. http://dx.doi.org/10.3390/su15054590

Jang, H, and L Soo-Bum. "Serving Robots: Management and Applications for Restaurant Business Sustainability." *Sustainability 12*, no. 10 (2020): 3998. http://dx.doi.org/10.3390/su12103998

Kohl, J L, M J Schoor, A M Syre, and D Gohlich. "Social Sustainability in the Development of Service Robots.". *Proceedings of the Design Society: DESIGN Conference 1*, (2020): 1949–1958. https://doi.org/10.3390/su12103998

Koyuncu, B, and M Guzel. "Software Development for the Kinematic Analysis of a Lynx 6 Robot Arm." *World Academy of Science, Engineering and Technology*, (2007): 252–257.

Manseur, R. "A Software Package for Computer-aided Robotics Education." *In Proceedings of 26th Annual Conference on Frontiers in Education 3*, (1996): 1409–1412. https://doi.org/10.1109/FIE.1996.568528

Mittal, R K, and I J Nagrath. 2003. *Robotics and Control*. 1st ed. New Delhi: Tata McGraw Hill Publishing Company Limited.

Niku, S B. 2001. *Introduction to Robotics: Analysis, Systems, Applications*. 4th ed. Prentice Hall.

Ogbemhe, J, M Khumbulani, and S T Nkgatho. "Achieving Sustainability in Manufacturing Using Robotic Methodologies." *Procedia Manufacturing 8*, (2017): 440–446. https://doi.org/10.1016/j.promfg.2017.02.056

Pan, M, T Lineer, W Pan, H Cheng, and T Bock. "A Framework of Indicators for Assessing Construction Automation and Robotics in the Sustainability Context." *Journal of Cleaner Production 182*, (2018): 82–95. https://doi.org/10.1016/j.jclepro.2018.02.053

Pellicciari, Marcello, A Avotins, K Bengtsson, G Bercelli, N Bey, B Lennarston, and D Meike. "AREUS—Innovative Hardware and Software for Sustainable Industrial Robotics." *IEEE International Conference on Automation Science and Engineering (CASE)*, (2015): 1325–1332. https://doi.org/10.1109/CoASE.2015.7294282

Pravak Manual, Model 1055, New Delhi, 2008. 1st ed. (2008) Pravak Cybernetics.

Project STAMINA, Sustainable and reliable robotics for part handling in manufacturing automation, in *Stamina – Robot*. (2013) Denmark.

Rinaldi, M, M Caterino, and M Fera. "Sustainability of Human-Robot Cooperative Configurations: Findings from a Case Study." *Computers & Industrial Engineering 182*, (2023): 1–13. https://doi.org/10.1016/j.cie.2023.109383

Saha, S K. 2008. *Introduction to Robotics*. 3rd ed. Tata McGraw Hill.

Santana, P F, B Jose, and C Luis. "Sustainable Robots for Humanitarian Demining." *International Journal of Advanced Robotic Systems 4*, no. 2 (2007): 207–218. https://doi.org/10.5772/5695

Schraft, R D, and A Merklinger. "Robots in Close Co-operations with Persons – A Contribution to Safety Sensors for Service Robots." *Flexible Automation and Intelligent Manufacturing*, (1996): 206–215. https://doi.org/10.1615/faim1996.220

Shehu, N, and A Nuhu. "The Role Of Automation And Robotics In Builtings For Sustainable Development." *Journal of Multi-disciplinary Engineering Science and Technology 6*, no. 2 (2019): 9557–9560.

Sheng, L, W Yiqing, C Qingwei, and H Weili. "A New Geometrical Method for the Inverse Kinematics of the Hyper Redundant Manipulators." *In IEEE International Conference on Robotic and Biomimetics (ROBIO'06)* (2005): 1356–1359. https://doi.org/10.1109/ROBIO.2006.340126

Soori, Mohsen, Arezoo Behrooz, and Roza Dastres. "Optimization of Energy Consumption in Industrial Robots, A Review." *Cognitive Robotics* (2022): 142–157. https://doi.org/10.1016/j.cogr.2023.05.003

Stroupe, A, A Okon, M Robinson, T Huntsberger, and H Aghzerian. "Sustainable Cooperative Robotic Technologies for Human and Robotic Outpost Infrastructure Construction and Maintenance." *Autonomous Robots 20*, no. 2 (2006): 113–123. https://doi.org/10.1007/s10514-006-5943-4

Vast, T, B Raucent, and F Petit. "A Case Study of Interactive Design of a Product and Its Assembly Line." *Proceedings of 7th International FAIM Conference, Flexible Automation and Intelligent Manufacturing* (1997): 776–786. https://doi.org/10.1007/978-94-011-5588-5_53

Wang, L, A Mohammad, X V Wang, and B Schimdt. "Energy-efficient Robot Applications towards Sustainable Manufacturing." *International Journal of Computer Integrated Manufacturing 31*, no. 8 (2018): 692–700. https://doi.org/10.1080/0951192X.2017.1379099

Weisz, J, Y Huang, F Lier, S Sethumadhavan, and P Allen. "Robobench: Towards Sustainable Robotics System Benchmarking." *IEEE International Conference on Robotics and Automation (ICRA)*, (2016): 3383–3389. https://doi.org/10.1109/ICRA.2016.7487514

Yahya, S, M Moghavvemi, and A F Mohamed. "Geometrical Approach of Planar Hyper-redundant Manipulators: Inverse Kinematics, Path Planning and Workspace." *Simulation Modeling Practice and Theory 19*, (2011): 406–422. https://doi.org/10.1016/j.simpat.2010.08.001

8 Enhancing Thermal Performance of Protective Textiles through Smart and Sustainable Fabric with Internal Structure Adjustments

Recent Breakthroughs and Developments

Jnanaranjan Acharya, Virendra Kumar, Dipankar Bhanja, and Rahul Dev Misra
National Institute of Technology, Silchar, India

8.1 INTRODUCTION

Firefighters are exposed to life-threatening conditions during their operational duties, including high-temperature environments characterized by flames, radiation, and steam. In a recent incident that transpired in the United States in 2018, National Fire Protection Association data recorded a staggering 58,250 reported injuries among firefighters. These injuries encompass a spectrum of issues, ranging from first- to third-degree skin burns, instances of cardiac arrest, and the psychological trauma they endure as a consequence of their routine work, which involves confronting a broad spectrum of exposure scenarios spanning from 2.5 to 84 kW/m^2. Consequently, there exists an urgent and global imperative for researchers to concentrate their efforts on advancing the efficacy and protective capabilities of firefighter turnout gear.

The development and production of advanced smart textiles for firefighters necessitates a multidisciplinary approach that takes into account external environmental factors, the human body's internal thermoregulation mechanisms, and the thermal and mechanical behaviour of protective textiles when exposed to high temperatures.

DOI: 10.1201/9781003349877-8

In a study conducted by van Rijswijk et al. (2009), a thermoplastic integrated anionic polyamide-6 composite fabric was investigated, and it was created through a vacuum infusion process. This fabric exhibited improved thermal performance due to the deactivation of the initiator and constraints stemming from the low in-plane permeability of the fibre textile during various transport processes. Nowadays firefighters utilize smart fabrics that possess a range of features, such as colour-changing capabilities, temperature-regulating properties, shape memory characteristics, waterproof and breathable attributes, and electronic information dissemination (Júnior et al., 2022). Cui and Liu (2021) innovatively designed an intelligent textile incorporating pneumatic actuators, often referred to as a "soft-logic fabric." This technology draws inspiration from the thermal regulation mechanisms observed in bird plumage, resulting in a fabric that exhibits a remarkable 15-fold increase in thermal resistance when inflated, delivering 3.84 times greater thermal comfort compared to conventional fabrics. Tu et al. (2022) conducted a comprehensive investigation into the effects of thickness, ambient temperature, and emissivity on various air-inflatable textiles and fibre-filled textiles. Their findings revealed that greater air thickness is not advisable for achieving elevated thermal insulation, and lower ambient temperatures do not significantly impact the thermal insulation of fibre-filled textiles. In another study by Shen et al. (2023), the influence of incorporating a windproof layer into firefighting fabrics was examined. The results showed that the thermal insulation of such fabrics remained largely unaffected during windy weather conditions, while the thermal insulation of porous protective fabrics experienced a substantial 211% alteration under windy conditions. Voirin (2015) experimentally conducted an investigation focused on the deployment of smart emergency attire, which seamlessly integrates sensor and communication technologies. This attire was designed to comprehensively monitor various thermophysiological parameters of firefighters, including heart rate, body temperature, and respiration rate, alongside environmental parameters encompassing location, chemical composition, thermal conditions, and altitude-related factors. Zavec et al. (2023) engineered a novel variant of protective garments, featuring the integration of nonwoven lightweight aerogels and an infrared-reflective textile layer. This innovative design resulted in a notable reduction in the thickness and overall weight of the garments, achieving a reduction of 15–20%, all while maintaining adherence to established work wear standards. Sousa et al. (2023) devised protective clothing embedded with an electromagnetic shielding capability, employing filler materials such as TiO_2/Fe_2O_3, TiO_2/ZnO, and multi-walled carbon nanotubes. This incorporation was achieved through the precise application of these materials via the knife-over-roll coating technique, yielding fabrics coated with TiO_2/Fe_2O_3 that demonstrated superior electromagnetic shielding efficacy exceeding 30 dB. Peng and Cui (2020) stratified smart protective clothing into two fundamental categories, specifically active cooling textiles (ACTs) and thermal-responsive textiles (TRTs). ACTs orchestrate internal thermal regulation by employing electronic cooling systems, whereas TRTs exhibit direct responsiveness to textile temperature in response to external perilous circumstances. Prior research has demonstrated the advantageous utilization of graphene electrodes (Diallo et al., 2022), which exhibit superior electrical and thermal conductivity, achieved through a process involving impregnation and reduction (Li et al., 2022; Feng et al., 2013) when applied to various fabric substrates.

The integration of graphene electrodes into protective textiles yields multiple benefits, including heightened thermal insulation and stability (Bhattacharjee et al., 2019; X. Zhang et al., 2023) through temperature reduction within the fabric. Furthermore, this integration enhances the capacity for sensing human motion and effectively ascertaining the presence of NO_2 gas (Luo et al., 2023).

The present inquiry is primarily focused on ongoing advancements within protective textiles, systematically addressing the following sectors:

1. Alterations to the internal fabric structures with the intent of augmenting the protective efficacy of firefighters' turnout gear.
2. The incorporation of smart coating into the protective textile to elevate its thermal performance metrics.
3. The integration of temperature-responsive shape memory materials into fire-resistant garments, ultimately heightening their operational capabilities.

8.2 CUTTING-EDGE INNOVATIONS IN PROTECTIVE TEXTILE MATERIALS

This section places a critical emphasis on the optimization of protective textile materials through the modification of internal structures, the application of intelligent coatings, and the incorporation of shape memory materials, while also providing an overview of sustainable fabric alternatives and the integration of digital technologies within the existing textile industry.

8.2.1 Modifications to the Internal Fabric Structures

Maurya et al. (2023) conducted a study examining the impact of process parameters, including the number of layers in the thermal liner, pick density of the outer shell, and variations in influx conditions on protection times. Their analysis, employing a Box-Behnken model, indicated that increased pick density has an inconsequential effect on protection time but does enhance the desired mechanical properties. Subham et al. (2023) made enhancements to the thermal liner of protective clothing by incorporating silica aerogel. Their findings revealed that an increase in aging and silylation time leads to improved protection against extreme heat loads. Notably, ammonium fluoride treatment yielded the best results in enhancing the properties of the aerogel in the Nomex thermal liner. Kumar et al. (2023) delved into the domain of moisture management and drying behaviour of the knitted inner layer within protective fabric. Their research demonstrated that a higher wool content contributes to improved moisture management and drying characteristics. Furthermore, knitted fabrics featuring a high filament count and a pile knit structure displayed superior performance for innerwear applications. In the investigation by Das et al. (2022), it was observed that thermal protective ratings decline across all aerial densities and numbers of layers. However, a significant impact of aerial density and the number of layers was particularly noticeable at lower incident radiation levels. Maurya et al. (2022) explored the efficacy of protective clothing with a single-layer rib-knitted structure in extending second-degree burn time under medium radiant exposure. This was achieved through the integration of a specific miss-stitch technique. Figure 8.1 depicts the different steps in fabric preparation.

8.2.2 Smart Coated Protective Fabric

A coating is applied to enhance and optimize the performance and technical attributes of fabrics. Smart coated techniques have been visually illustrated in Figure 8.2. Texfire coatings are formulated by combining a flame-resistant textile foundation.

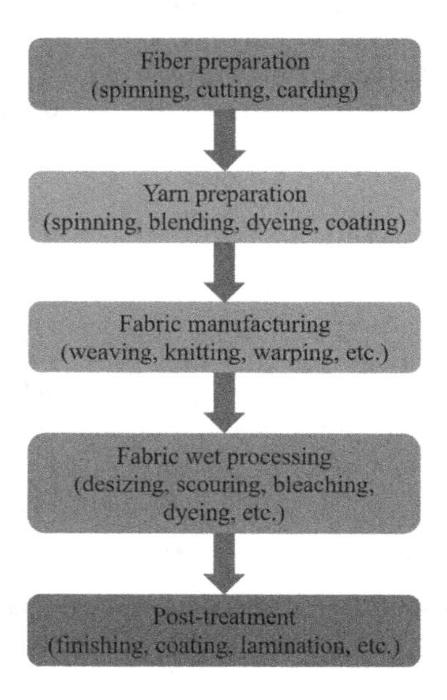

FIGURE 8.1 Steps in fabric preparation process.

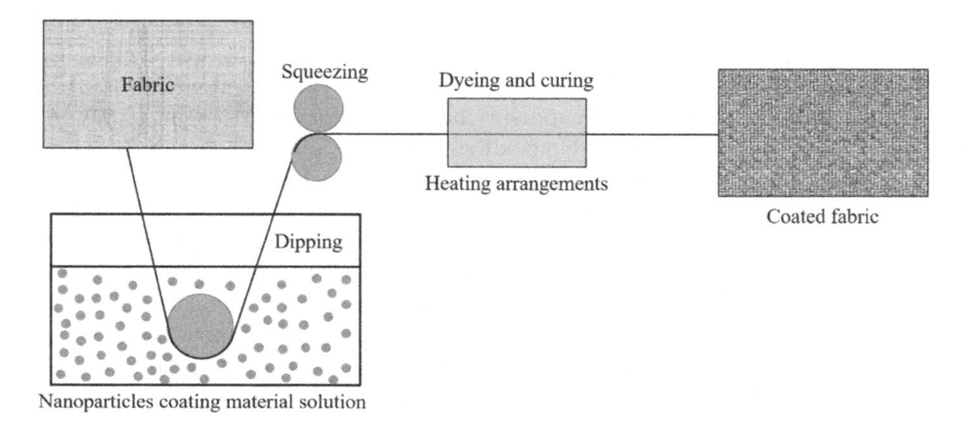

FIGURE 8.2 Techniques employed in the fabric coating procedure.

These coatings can be applied to either one or both sides of the fabrics, depending on their intended use. Miedzińska et al. (2021) conducted a study in which they applied a reflective Ti-Si-N nanocomposite coating to fire-resistant fabrics, specifically NATAN and PROTON (trademarked names). They employed magnetron sputtering techniques for the coating process and evaluated the increase in temperature and the time required to reach the pain threshold, known as a second-degree burn, according to the Henrique burn criterion. The inclusion of a Ti-Si coating notably enhanced the heat resistance of both PROTON and NATAN fabrics across all tested heat intensity levels. For instance, in the case of PROTON fabric, the time needed to reach 60°C increased from 11.23 seconds (without the coating) to 13.13 seconds (with the Ti-Si coating) at a heat level of 0.615 kW/m^2. As for NATAN fabric, the increase was most pronounced, rising from 7.76 seconds (without the coating) to 11.30 seconds (with the Ti-Si coating) at the same heat level. Manasoglu and Kanik (2022) investigated the impact of coating polyester woven fabrics with micronized perlite particles of varying sizes (10–38, 50–63, 100–150 m) and concentrations (20, 40, 60, 80 g/kg) using the knife-over-roll method. Their research focused on the solar and thermal insulation properties of these coated fabrics. The study concluded that the heat resistance properties of perlite-coated fabrics were significantly enhanced. Specifically, the lowest thermal conductivity and heat transmission factor (0.088 W/mK and 26.83%) were achieved with larger perlite particles at higher concentrations. Consequently, increasing the concentration of micronized perlite particles, regardless of their size, led to a decrease in solar transmittance and an increase in solar reflectance. Prasad et al. (2018) worked on rejuvenating a highly reflective coating applied to fabric substrates to enhance the reflective properties of turnout gear. Highly reflective coatings involve the deposition of a thin film of coating materials onto the substrate. These coatings come in two varieties: those composed of metal and those made of dielectric materials. Research on composite coatings (Morgiel et al., 2006, 2007; Dąbrowski et al., 2009), such as titanium nitride-silicon nitride and chromium nitride-silicon nitride, has provided valuable insights into coating application and the influence of application process conditions on the structure of these coatings. This research has expanded understanding regarding the interplay between coating properties and the observed structural outcomes. Phase change materials (PCMs) have the capability to store or release heat within a specific temperature range. These materials can undergo a change in state over a narrow temperature span. During the heating process, they absorb energy and change state, while in the reverse cooling process, they release energy to the external environment. PCMs act as thermal insulators, and the efficacy of their insulation varies with temperature and time. Researchers have recently shown an increased interest in incorporating PCMs into textiles, either through coating or encapsulation, in order to create intelligent textiles capable of regulating temperature. Furthermore, there is a growing interest in fibers and fabrics that can respond to changing temperatures on their own. This adaptability is achievable through the utilization of PCMs (Weder, 2001). The technology of incorporating microcapsules of PCM (Nelson, 2001) into textile structures to enhance their temperature-regulating capabilities originated in the early 1980s as part of a NASA research initiative. Initially, this innovation was aimed at creating fabrics for astronauts' space suits, enhancing thermal protection against the extreme temperature fluctuations experienced in outer space. In the context of a firefighter's turnout gear, the thermal liner plays a pivotal role in improving thermal protection performance (Zhang et al., 2017). Researchers employed a modified thermal protective performance tester to investigate the impact of adding aerogel and microencapsulated phase change materials to thermal liners, which typically comprise a traditional thermal liner, a phase-change layer, and an aerogel layer. The study also delved into the crucial factors contributing to the enhanced effectiveness of these thermal liners.

8.2.3 Temperature Responsive Material Integrated Fabric

The development of an alloy endowed with the ability to respond to environmental conditions and exhibit self-adjusting properties is of paramount significance in elevating the thermal protective performance of protective turnout gear. This advancement serves to significantly reduce the likelihood of skin burn injuries among firefighters and emergency rescuers. Shape memory materials stand out as intelligent materials, possessing the unique capacity to revert to their original shape when subjected to specific environmental stimuli following substantial deformation, as shown in Figure 8.3. Incorporating shape memory alloys (SMAs) into firefighters' protective turnout gear results in a substantial augmentation of protective performance. These SMAs operate effectively and efficiently, thereby mitigating the risks to the lives

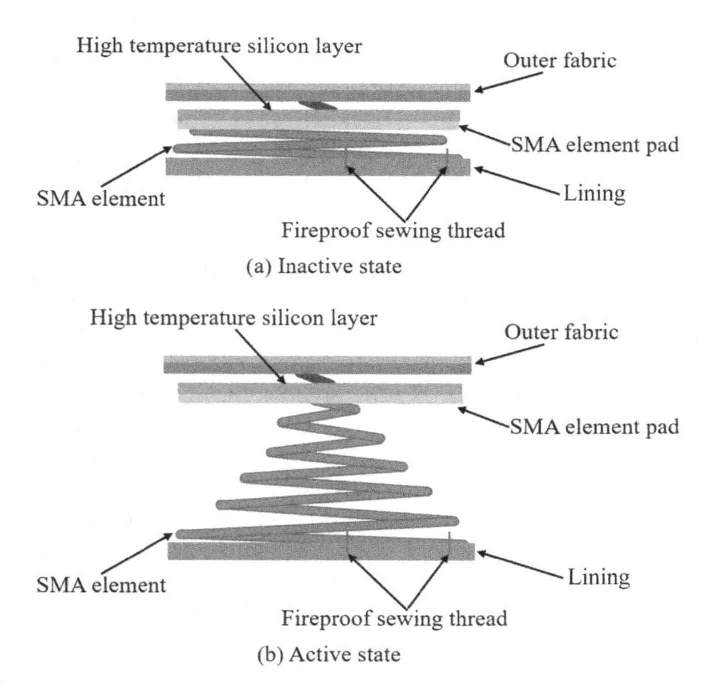

FIGURE 8.3 Fabric incorporating shape memory alloy (SMA) in (a) inactive and (b) activated configurations.

and physiological well-being of these professionals (Ma et al., 2019). Wang et al. (2022) pioneered the development of a shape memory fabric by combining NiTi filaments with aramid yarns, interspersed at varying distances of 2, 3, and 4 cm between adjacent filaments. This specialized fabric, in its activated state, demonstrates superior heat insulation properties and offers reduced resistance to moisture evaporation compared to conventional three-layer fabrics. Researchers have introduced an innovative approach by integrating SMA components into the turnout gear worn by firefighters. The shape memory alloy in a fabric system can manifest in two forms. In the initial configuration, a two-way SMA wire (Congalton, 1999) adopts a spring-like shape at elevated temperatures and remains flat at martensitic temperatures. This wire is integrated into a three-layer fabric system. Upon exposure to heat, the spring-shaped alloy reverts to its original form, creating an augmented layer of insulating air between the fabric's layers. In an alternative approach, the two-way SMA (Yates, 2012) wire is subjected to heating, causing it to assume a central arched eight-shaped loop, which is then cooled and flattened. This wire is affixed to the pockets of firefighters' attire. While both methods serve to reduce the risk of skin burns, subsequent research has favoured the first alloy shape due to its more consistent maintenance of shape.

8.2.4 Advancements in Sustainable and Technologically Enhanced Protective Fabrics

Nie et al. (2023) have introduced an innovative, environmentally friendly, and biodegradable thermosensitive ink with advanced properties. This ink is created by employing a triblock copolymer, specifically PCLA-PEG-PCLA, which eliminates the need for chemical pre-treatment processes. The ink effectively reduces shrinkage during heating by 41% and 51% along the weft and warp directions, respectively. Halbrecht et al. (2023) have designed a 3D-printed knitted spacer fabric that is non-uniform and breathable, offering enhanced compression strength and adaptable suitability for a range of protective applications. The fields of materials science, polymer science, nanoscience, and applied physics have all paid close attention to the development of ionic polymer-metal composites because of their unique combination of properties, which include high flexibility, large bending and deformation, good water absorption, and biocompatibility (Biswal et al., 2022; Wang et al., 2023). Cheng et al. (2022) have successfully engineered a high-performance, environmentally sustainable bilayer polypropylene-metalized polypropylene (PPM) material. This innovative PPM material is produced by sputtering cotton with Ag/Zn and incorporating PLA/PCL cellulose nanofibers. This material exhibits superior antibacterial properties, effectively restricts the movement of particulate matter, and efficiently manages the exhaled water vapour from human sources. He et al. (He et al., 2021) employed a two-dimensional digital image correlation technique to scrutinize the strain components in play at the crack tip when subjecting PVC coated fabrics to biaxial loading conditions. Their findings indicate that the tearing residual strength (TRS) remains relatively consistent when the crack orientation is altered for a fixed initial crack length. However, the maximum TRS is observed when the crack is oriented at 45°. Ali et al. (2022) employed deep convolutional neural networks to perform segmentation on micro-computed tomography (μCT) images of a multilayer plain-woven fabric. Subsequently, individual yarns are distinguished from interconnected yarns using the watershed segmentation technique, contributing to the creation of digital material replicas.

8.3 CONCLUSION

The study offers a comprehensive exploration of recent advancements in protective clothing, with a focus on enhancing its performance in the face of diverse hazardous environments. The present review delves into internal fabric modifications designed to augment protection times while simultaneously reducing fabric temperature. It also thoroughly explores smart coatings, including reflective and phase change material coatings, integrated into turnout gear, shedding light on their impact on mitigating thermal damage. Furthermore, the analysis provides insights into the utilization of temperature-responsive materials, specifically shape memory materials, strategically incorporated within the fabric ensemble to alleviate the severity of burn injuries. This review serves as a valuable resource for researchers, facilitating the pursuit of innovative approaches in the development of high-performance smart protective clothing, ultimately enhancing the safety and well-being of firefighters.

REFERENCES

Ali, M. A., Guan, Q., Umer, R., & Cantwell, W. J. (2022). Efficient processing of μ CT images using deep learning tools for generating digital material twins of woven fabrics. *Composites Science and Technology*, *217*, 109091. https://doi.org/10.1016/j.compscitech.2021.109091

Bhattacharjee, Shovon, Joshi, R., Chughtai, A. A., & Macintyre, R. C. (2019). Graphene modified multifunctional personal protective clothing. *Advanced Materials Interfaces*, *6*(21), 1–27. https://doi.org/10.1002/admi.201900622

Biswal, D.K., Moharana, B.R. and Mohapatra, T.P., 2022. Bending response optimization of an ionic polymer-metal composite actuator using orthogonal array method. *Materials Today: Proceedings*, *49*, pp.1550–1555.

Cheng, Y., Li, J., Chen, M., Zhang, S., He, R., & Wang, N. (2022). Environmentally friendly and antimicrobial bilayer structured fabrics with integrated interception and sterilization for personal protective mask. *Separation and Purification Technology*, *294*(April), 121165. https://doi.org/10.1016/j.seppur.2022.121165

Congalton, D. (1999). Shape memory alloys for use in thermally activated clothing, protection against flame and heat. *Fire and Materials*, *23*(5), 223–226. https://doi.org/10.1002/(SICI)1099-1018(199909/10)23:5<223::AID-FAM687>3.0.CO;2-K

Cui, Y., & Liu, X. (2021). Soft-logic: design and thermal-comfort evaluation of smart thermoregulatory fabric with pneumatic actuators. *Journal of the Textile Institute*, *112*(12), 1913–1924. https://doi.org/10.1080/00405000.2020.1848121

Dąbrowski, M., Morgiel, J., Grzonka, J., Mania, R., & Zimowski, S. (2009). N/TiN coatings on sintered carbide turning inserts. *Elektronika*, *50*, 71–73.

Das, T., Das, A., & Alagirusamy, R. (2022). Study on thermal protective performance of thermal liner in a multi-layer clothing under radiant heat exposure. *Journal of Industrial Textiles*, *51*(5_suppl), 8208S–8226S. https://doi.org/10.1177/15280837221094057

Diallo, A. K., Helal, E., Gutiérrez, G., Madinehei, M., David, É., Demarquette, N., & Moghimian, N. (2022). Graphene: A multifunctional additive for sustainability. *Sustainable Materials and Technologies*, *33*(September). https://doi.org/10.1016/j.susmat.2022.e00487

Feng, Q. Q., Zhu, F. L., & Yang, K. (2013). The application of outlast acrylic temperature controlling fabric in firefighters protective clothing. *China Personal Protective Equipment, 4*(004), 39–44.

Halbrecht, A., Kinsbursky, M., Poranne, R., & Sterman, Y. (2023). 3D printed spacer fabrics. *Additive Manufacturing, 65*(August 2022), 103436. https://doi.org/10.1016/j.addma.2023.103436

He, R., Sun, X., Wu, Y., Tang, G., & Carvelli, V. (2021). Biaxial tearing properties of woven coated fabrics using digital image correlation. *Composite Structures, 272*(May), 114206. https://doi.org/10.1016/j.compstruct.2021.114206

Júnior, H. L. O., Neves, R. M., Monticeli, F. M., & Dall Agnol, L. (2022). Smart fabric textiles: Recent advances and challenges. *Textiles, 2*(4), 582–605. https://doi.org/10.3390/textiles2040034

Kumar, K., Das, A., Vishnoi, P., & Singh, J. P. (2023). Moisture management properties and drying behaviour of knitted fabrics for inner layer applications. *Indian Journal of Fibre and Textile Research, 48*(1), 91–98. https://doi.org/10.56042/ijftr.v48i1.59978

Li, R. K., Li, R. C., & Zhu, L. (2022). System of seven-lead electrocar- diogram monitoring based on graphene fabric electrodes. *Journal of Textile Research, 43*(7), 149–154.

Luo, Y., Miao, Y., Wang, H., Dong, K., Hou, L., Xu, Y., Chen, W., Zhang, Y., Zhang, Y., & Fan, W. (2023). Laser-induced Janus graphene/poly(p-phenylene benzobisoxazole) fabrics with intrinsic flame retardancy as flexible sensors and breathable electrodes for firefighting field. *Nano Research, 16*(5), 7600–7608.

Ma, N., Lu, Y., He, J., & Dai, H. (2019). Application of shape memory materials in protective clothing: A review. *Journal of the Textile Institute, 110*(6), 950–958. https://doi.org/10.1080/00405000.2018.1532783

Manasoglu, G., & Kanik, M. (2022). Investigation of thermal and solar properties of perlite coated woven fabrics. *Journal of Applied Polymer Science, 139*(4), 1–15. https://doi.org/10.1002/app.51543

Maurya, S. K., Uttamrao Somkuwar, V., Garg, H., Das, A., & Kumar, B. (2022). Thermal protective performance of single-layer rib-knitted structure and its derivatives under radiant heat flux. *Journal of Industrial Textiles, 51*(5_suppl), 8865S–8883S. https://doi.org/10.1177/15280837211042680

Maurya, S., Rathour, R., Mishra, P., Mishra, D., Das, A., & Ramasamy, A. (2023). Studies on the effect of process parameters on the thermal protective performance of multilayer thermal protective clothing. *Journal of the Textile Institute, 0*(0), 1–12. https://doi.org/10.1080/00405000.2023.2250486

Miedzińska, D., Giełżecki, J., Mania, R., Marszalek, K., & Wolański, R. (2021). Influence of ti-si-n nanocomposite coating on heat radiation resistance of fireproof fabrics. *Materials, 14*(13). https://doi.org/10.3390/ma14133493

Morgiel, J., Major, Ł., Grzonka, J., Mania, R., & Rakowski, M. (2006). Elaboration of magnetron deposition conditions of Tin/Si 3 N 4 nanocomposite coatings. *Problemy Eksploatacji, 2*, 33–42.

Morgiel, J., Mania, R., Grzonka, J., Rogal, Ł., Janus, A. M., & Zientara, D. (2007). Consolidating condition of Cr-Si compacts and their microstructure. *Archives of Materials Science and Engineering, 11*(28), 673–676.

Nelson, G. (2001). Microencapsulation in textile finishing. *Review of Progress in Coloration and Related Topics, 31*, 57–64.

Nie, L., Chen, Y., Dong, Y., Li, R., & Chang, G. (2023). Colloids and surfaces A: Physicochemical and Engineering Aspects Development of high-performance, chemical pretreatment-free dye-based inks for digital printing on polyester fabric. *Colloids and Surfaces A: Physicochemical and Engineering Aspects, 678*(September), 132470. https://doi.org/10.1016/j.colsurfa.2023.132470

Peng, Y., & Cui, Y. (2020). Advanced textiles for personal thermal management and energy. *Joule, 4*(4), 724–742. https://doi.org/10.1016/j.joule.2020.02.011

Prasad, K., Goyal, A., Gohil, K., & Jagyasi. (2018). Highly reflective coatings. *International Journal of Applied Engineering Research, 13*(22), 15773–15782.

van Rijswijk, K., Teuwen, J. J. E., Bersee, H. E. N., & Beukers, A. (2009). Textile fiber-reinforced anionic polyamide-6 composites. Part I: The vacuum infusion process. *Composites Part A: Applied Science and Manufacturing, 40*(1), 1–10. https://doi.org/10.1016/j.compositesa.2008.03.018

Shen, H., Teng, F., Bai, T., Zhang, G., Chen, Y., Wen, R., Xu, G., Wang, M., Wang, F., Wang, J., & Tu, L. (2023). Influence of windproof layer on heat and mass transfer in porous textiles based on experiment and simulation. *International Communications in Heat and Mass Transfer, 142*(February). https://doi.org/10.1016/j.icheatmasstransfer.2023.106667

Sousa, A., Barbosa, J., Soares, O., Ferreira, J., Gonçalves, A., Santos, G., Silva, A., & Morgado, J. (2023). Tailored materials for electromagnetic shielding textile application. *In 10 Th European Conference on Protective Clothing*, 72.

Subham, R., Das, A., & Alagirusamy, R. (2023). Development of thermal liner for extreme heat protective clothing using aerogel technology. *Journal of the Textile Institute, 0*(0), 1–10. https://doi.org/10.1080/00405000.2023.2201913

Tu, L., Zhang, H., Wang, J., Sun, J., Xu, G., Wang, F., & Shen, H. (2022). Numerical investigation of heat transfer and fluid motion in air inflatable textiles: Effect of thickness, surface emissivity and ambient temperature. *Building and Environment, 225*(September), 109594. https://doi.org/10.1016/j.buildenv.2022.109594

Voirin, G. (2015). Working garment integrating sensor applications developed within the PROeTEX project for firefighters. Ubiquitous Computing in the Workplace: What Ethical Issues? In *An Interdisciplinary Perspective* (pp. 25–33).

Wang, L., Pan, M., Lu, Y., Song, W., Liu, S., & Lv, J. (2022). Developing smart fabric systems with shape memory layer for improved thermal protection and thermal comfort. *Materials and Design, 221*, 110922. https://doi.org/10.1016/j.matdes.2022.110922

Wang, H., Yang, L., Yang, Y., Zhang, D., & Tian, A. (2023). Highly flexible, large-deformation ionic polymer metal composites for artificial muscles: Fabrication, properties, applications, and prospects. *Chemical Engineering Journal, 469*, 143976.

Weder, M. (2001, October). Scale of change. *Textile Month*, 37–38.

Yates, D. A. (2012). *Design and Evaluation of a Thermally Responsive Firefighter Turnout Coat*. University of Maryland, College Park.

Zavec, D., Richter, K., Lachmann, E., Cherunova, I., Kornev, N., Jia, G., & Plentz, J. (2023). ArTiShirt–workwear for cold and arctic temperatures. *In 10th European Conference on Protective Clothing*, 22.

Zhang, H., Song, G., Su, H., Ren, H., & Cao, J. (2017). An exploration of enhancing thermal protective clothing performance by incorporating aerogel and phase change materials. *Fire and Materials, 41*(8), 953–963. https://doi.org/10.1002/fam.2435

Zhang, X., Tian, M., & Li, J. (2023). Influence of smart textiles on the thermal protection performance of firefighters' clothing: A review. *The Journal of The Textile Institute*, 1–14.

9 Role of Industrial Internet of Things (IIoT) Enabled Smart Warehousing for Sustainable Perishable Food Supply Chains

Arpita Nayak
KIIT University, Bhubaneswar, India

Atmika Patnaik
King's College, London, United Kingdom

Ipseeta Satpathy, Sushanta Tripathy, and B.C.M. Patnaik
KIIT University, Bhubaneswar, India

9.1 INTRODUCTION

The Internet of Things (IoT) is a network of intelligent devices that gather, analyze, and act on data in real-time, therefore enhancing processes in industry and manufacturing. The Industrial Internet of Things (IIoT) refers to the use of IoT technologies in manufacturing. Industry 4.0 is another name for IIoT, which enables real-time analytics and decision-making by using data created by traditional industrial machines over time. The core idea behind IIoT is that smart devices are superior to humans in terms of gathering and evaluating data in real time, as well as delivering critical information quickly. This data may be utilized to make faster and more accurate business decisions. For example, India's National Horticulture Mission anticipates that more than US $8.5 billion will be deployed in warehousing in 2019, with 101 high-tech cold storage units approved.

IIoT is an advanced network that connects intelligent devices to create complete systems. These systems are designed to monitor, gather, share, and analyze data in a range of industrial settings. Each IIoT ecosystem is made up of sensors that can detect, communicate, and store details on themselves. A reliable public or private

DOI: 10.1201/9781003349877-9

data communications infrastructure is also necessary to allow seamless data flow. Analytics and apps, on the other hand, play an important role in analyzing the raw data provided by these networked devices and translating it into useful business insights. Adequate storage solutions are also required to handle the massive amounts of data created by IIoT devices. Finally, human engagement is critical since people are essential to monitoring and using the actionable information received from the IIoT network. This vital data may be used to improve predictive maintenance and overall company operations. The infrastructure of IIoT is described in Figure 9.1 (Posey et al., 2022; Boyes et al., 2018).

The Internet of Things has surely contributed to the Fourth Industrial Revolution. Inbound and outbound management operations are usual in a warehouse. On the one hand, the inbound process includes actions such as order receipt and restocking, with the goal of receiving items and storing things until they are needed. The outbound method, on the other hand, reacts directly to client expectations and involves duties such as order selection, packaging, and delivery.

Order picking is the most labor-intensive function involved in order handling at a distribution center, accounting for 55% of total expenditures (Accorsi et al., 2014). Robots have been used to automate logistics warehousing systems thanks to developments in wireless communication and embedded computers. Companies like Amazon, DHL, and Alibaba, for example, use a huge number of automated guided

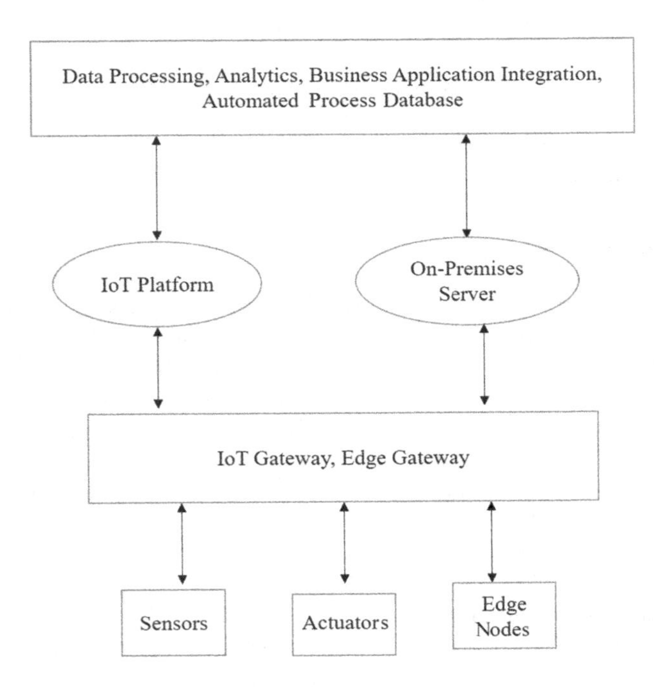

FIGURE 9.1 Infrastructure of IIoT.

***Source*: Adopted from Posey et al., 2022**

vehicles to improve the efficiency of their operations (Lee, 2018). Manufacturers are rapidly embracing the IoT in conjunction with newer blockchain technologies to achieve a higher degree of automation, accountability, safety, and efficiency. According to Data Bridge Market Research (2023), the well-established worldwide IIoT platform market is predicted to increase from $394.48 billion in 2022 to $1,809.04 billion by 2030, reflecting an annualized annual growth rate (CAGR) of 20.97%. Meanwhile, according to Research and Markets, the nascent worldwide market for blockchain in manufacturing will expand at a CAGR of 72% from $1.2 billion in 2022 to $88.5 billion by 2030. Business-to-business IoT apps will generate $300 billion in revenue by 2022. According to comparable McKinsey research, the advantages of IoT in supply chain management (SCM), inventory management, and logistics might total $560 billion by 2025. This is unsurprising considering that the IoT is expanding to consist of devices such as headphones, smartphones, wearables, and lighting.

Errors in the warehouse may be quite costly. The biggest issue for a warehouse manager in a typical structure is the lack of data to make an educated choice. Incorrect procedures need more labor to fix, raising overall expenses. As a result, businesses are already incorporating IoT sensors into their warehouses to track the movement of products as well as the utilization of resources and other assets in manufacturing units. They also employ sensors on shelves to communicate immediate inventory information to their management tool, which saves money and time (Sharma & Villányi, 2022). A successful IoT network enables remote monitoring and administration of the whole warehouse, saving money and time on duties like inventory identification, control, and stock movement tracking. Warehouse management effectiveness will be greatly improved by using wireless internet, networks of sensors that collect real-time data, and IoT platforms.

The methods and procedures implicated in the shipment, preservation, and distribution of perishable food products while minimizing environmental effects, decreasing waste, and preserving the food supply's long-term sustainability are referred to as sustainable perishable food supply chains. These supply chains seek to balance all three of these variables in order to establish a resilient and sustainable food system. Perishable items have particular handling procedures that may have impacts on society and the environment in addition to the well-known economic considerations. Food-sustainable supply chain management is becoming increasingly significant (Jouzdani & Govindan, 2021). Food supply networks differ from those for other goods due to intrinsic characteristics, most notably significant and constant changes in food product quality across the supply chain (Soysal & Çimen, 2017). If a product fit one of the following requirements, it is deemed ephemeral: 1) It suffers a major drop in quality or quantity, 2) the loss of functioning has serious repercussions, or 3) its worth falls over time. Food goods, in particular, are considered volatile since their quality degrades fast not only during the manufacturing and storage procedures but also throughout distribution. The capacity of food supply networks to deliver fresh foods to clients is critical to their viability and profitability. In this context, the architecture of a perishable food supply chain, like any other sort of supply chain, is important (Musavi & Bozorgi-Amiri, 2017). The adverse effect of food loss and waste (FLW) for each food group was assessed by the author. FLW was accounted

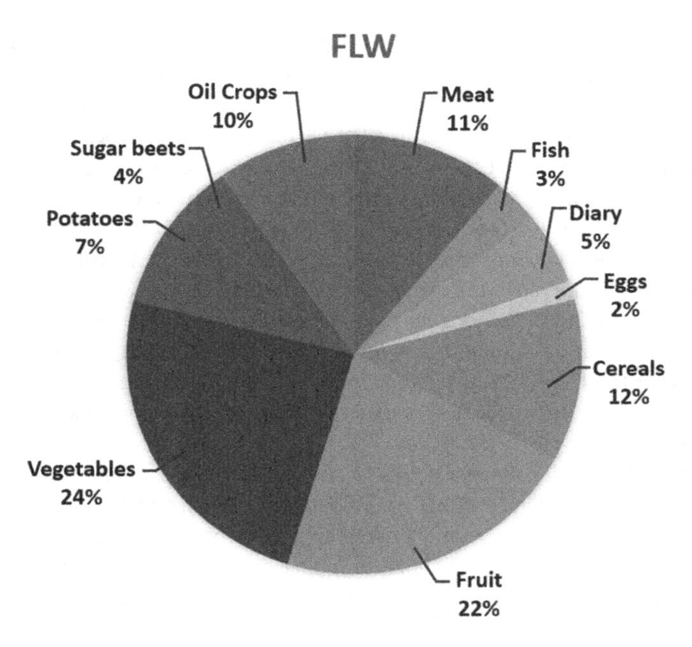

FIGURE 9.2 FLW of each food group.

***Source*: Adopted from De Boni et al., 2022**

for by vegetables and fruits (24%), cereals (12%), oil crops (10%), potatoes (7.3%), and sugar beets (4%). Animal-origin foods provided the least amount of FLW, including meat (11%), dairy (5.3%), fish (3.3%), and eggs (1.4%) as shown in Figure 9.2.

Furthermore, cutting-edge technologies like IoT sensors, blockchain, and data analytics are being used to improve traceability, openness, and efficiency in perishable food supply chains. Moisture, temperature, and other elements that impact food quality may be monitored in real-time using these technologies. Supply chain stakeholders may identify areas for enhancement, optimize storage and transportation circumstances, and make educated decisions to decrease waste and increase sustainability by collecting and analyzing data (Jabbar et al., 2018). The various benefits of IIoT in smart warehousing for sustainable perishable food supply chains are mentioned as follows (Evtodieva et al., 2020; Pal & Kant, 2020):

- Smart energy management in warehouses is made possible by IIoT solutions, which optimize the use of lighting, cooling, and ventilating systems depending on real-time demand and occupancy. This helps to cut energy usage and operating expenses while also guaranteeing that perishable food products are stored in optimal conditions. Furthermore, IIoT analytics may give insights into resource utilization, enabling greater resource optimization in areas like water, packaging components, and transit, resulting in increased sustainability.

- IIoT technology like automated guided vehicles, robots, and smart conveyors help to simplify warehouse operations and eliminate the need for manual labor. Automated methods allow for speedier order fulfillment, picking, and packaging, resulting in less handling time and more operational efficiency. In the perishable food supply chain, this leads to increased production, cost reductions, and shorter lead times.
- Temperature conditions may be monitored and controlled by IIoT devices at many phases of the cold chain, including transit and storage. This keeps perishable food products within the necessary temperature range, preventing spoiling. Real-time temperature monitoring, in conjunction with automatic warnings and notifications, enables prompt response in the event of any deviations, lowering the risk of quality degradation and assuring food safety.

To be ecologically sustainable, a corporation must be mindful of waste and pollution, both of which must be minimized. One of the most significant advantages of an IIoT-enabled smart warehouse to sustainable development is its capacity to deploy resources efficiently. A smart warehouse is capable of accurately tracking environmental elements such as temperature, humidity, light, and so on using sensors and real-time analytics. This control system enables energy transmission, which decreases utilization, waste, and the life of perishable products. This minimizes not just the emissions related to refrigeration and climate regulation, but also the environmental effect of food waste, which is a major global problem. Food waste is a major issue with far-reaching economic, environmental, and social consequences. Smart warehouses powered by IIoT play an essential role in decreasing food waste in the supply chain (Buntak et al., 2019). The ability to see inventory levels and conditions in real-time, including freshness and quality, allows for more precise demand forecasts and inventory management. This precision guarantees that perishables are transported and delivered effectively, eliminating the possibility of overpacking or damage. As a result, it helps sustainable development by preserving important resources, lowering greenhouse gas emissions associated with decaying food, and fulfilling the ethical duty of offering all people cultivating food to the globe.

Transparency and accountability, two fundamental components of sustainable development practices, are promoted by IIoT technology. Stakeholders in the supply chain may get vital information on product origins, quality, and consumption practices by using real-time data and analytics. This knowledge also enables a more complete examination of items derived from ethical and sustainable practices. It instills trust in consumers, who are increasingly seeking greater transparency into their goods, particularly in the food industry. In accordance with this expectation, IIoT-enabled smart warehouses promote sustainable and responsible sourcing (He et al., 2018; Chhotaray et al., 2023).

Advanced machine learning algorithms can also optimize inventory levels, minimize food waste, and combat overproduction. Supply chain optimization may reduce greenhouse gas emissions and transportation costs by streamlining routes and transportation. Waste monitoring systems that use AI-powered picture recognition to identify broken or ruined food can help reduce waste. These attempts can be aided

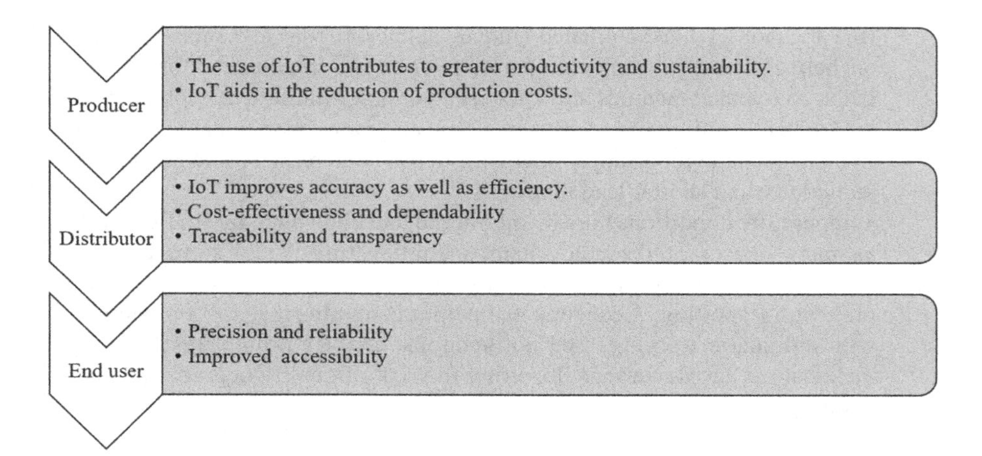

FIGURE 9.3 Using modern technology such as the internet of things (IoT), digital networks, robots, and cognitive data analytics to improve the efficacy and versatility of the agri-food supply chain.

***Source*: Adopted from Tan et al., 2023**

further by robots equipped with computer vision. One example of green supply chain optimization is solar panels and wind turbines, which promote responsible energy consumption by balancing energy supply and demand (Mehroof, 2019). Advanced technologies like IIoT and blockchain technology improve visibility in SCM, where a 'block' is an information package containing all preliminary information as well as fresh data. The use of this technology can help alleviate the problem of openness and security in stock management, boost operational reliability, secure confidential transactions and information availability, and reduce relationship timing between supply chain participants because of the change in management framework; where decisions are decentralized, blockchain technology decreases and minimizes the material flow parameter. Connectivity of technology-oriented digital networks, robots, and sophisticated data analytics with physical processes is critical to improving supply chain efficiency and effectiveness as depicted in Figure 9.3 (Tan et al., 2023).

9.1.1 Research Questions

In accordance with the scope of the study, a subsequent research question has been developed:

- How can IIoT-enabled smart warehousing affect real-time monitoring, predictive analytics, and inventory management in perishable food supply chains?
- What is the implication of IIoT-enabled smart warehousing towards sustainable development practices?

9.2 LEVERAGING THE POWER OF REAL-TIME MONITORING OF PERISHABLE FOOD PRODUCTS: THE APPLICATION OF IIOT-ENABLED WAREHOUSES

In the changing environment of perishable food supply chains, real-time monitoring is crucial for guaranteeing the safety, effectiveness, and quality of perishable food items. Warehouses have grown into intelligent and connected settings as a result of the advent of IIoT technology, enabling effective real-time monitoring of perishable food commodities.

For farmers and suppliers working with fresh or frozen perishable items, ensuring maximum quality and safety along the whole food cold chain is critical. The cold chain trip involves multiple hazards and can last several days or even weeks, involving operations like harvesting, processing, shipping, cold storage, and shipment to grocers and restaurants. Not only are there potentially many steps along this path, but there are also several components that might reduce the shelf life of your food, deteriorate its quality, or render it hazardous to eat.

Dairy, medicines, meat, flowers, and all perishable commodities have one thing in common: they are all checked during transportation to ensure freshness and usability (Salama et al., 2019). Because it provides vital information for evaluating food quality and predicting shelf life, environmental monitoring is essential for the management of the perishable food supply chain. To be marketed, all forms of perishable items must be given in high grade and quality condition. Proper storage conditions are required for this. Temperature and humidity are examples of such conditions, as is room access management. Certain food categories' quality diminishes with time owing to natural chemical processes, which can be slowed by using lower temperatures. Temperature monitoring can help with this. For perishable food safety and quality, real-time supply chain management is critical. Temperature mismanagement during perishable food shipment, for example, can result in quality degradation and up to 35% product loss (Wang et al., 2015). The EU 1169 food management regulation (rule (EU) No 1169/2011) was established in 2011 by the European Parliament and the Council of the European Union. Customers must get information on the influence of food on their health and safe food handling by December 2014, according to this guideline. Perishable food's quality and safety are strongly reliant on its resilience to the surroundings during shipping and storage. As a result, developing an IIoT-enabled real-time tracking device to gather environmental data along the supply chain is critical for assuring the safety and efficacy of perishable food items (Aung & Chang, 2014; Caldeira et al., 2019). Thermometers or humidity sensors are fitted in trucks and warehouses in conventional cold chain management systems (Alfian et al., 2017; da Costa et al., 2022). The diagram (see Figure 9.4) displays the three key components of IoT's contribution to the continuous surveillance of perishable food supply chains (Verdouw et al., 2016).

Control TCS – To prevent bacterial development and food spoiling, every food-related organization must maintain the right time and temperature for temperature-controlled for safety (TCS) food. Companies may improve their monitoring and tracking of TCS food temperature, humidity, and

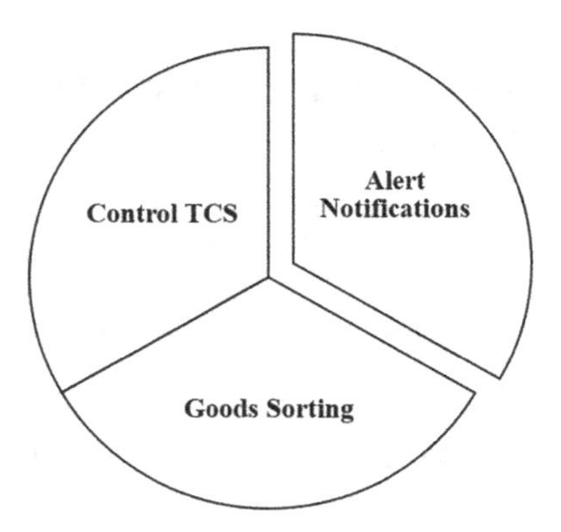

FIGURE 9.4 Major benefits of IIoT in real-time monitoring of perishable food products.

***Source*: Author's own compilation**

other essential factors by implementing an IoT remote monitoring system. Organizations are able to monitor the temperature of the food, ensuring it remains in the right environment, and make any modifications by integrating IoT solutions into their vehicle fleets and storage systems.

Alert Notifications – When there are temperature control violations, quick action is required to prevent food wastage in perishable items. IoT solutions play an important part in this by delivering quick push notifications to key staff in the case of cold chain management problems.

Real-time alerts emphasize temperature dips in cooling systems, vehicle faults, and other difficulties that may develop during food distribution. This helps the cold chain management team to quickly identify and fix the issue, minimizing perishable food loss.

Good Sorting – Initially, organizations hand-sort perishable items based on ripening stages for sales. Now, IoT technology efficiently assists in sorting TCS goods without human mistakes, which saves both time and money. Most grocery businesses benefit from employing IoT technology to preserve food quality and freshness, which is a major issue among consumers.

A network of sensors and equipment in IIoT-enabled warehouses continually monitors important environmental characteristics such as temperature, humidity, and air quality. These sensors produce real-time data, allowing for the quick identification of deviations from ideal circumstances. Warehouse operators and workers receive timely warnings and notifications, allowing them to take rapid corrective steps to keep perishable food goods in the proper storage climate (Kumar et al., 2021; Galgal et al., 2023). Real-time data from IIoT-enabled warehouses are used to generate alarms and warnings in the event of important occurrences or departures from established criteria. If the temperature in a cold storage facility, for example, increases over an allowed level, an instant alarm is issued to approved employees. This timely notice allows for a quick response and remedial action to limit any potential harm that might occur to perishable food supplies (Sverko et al., 2022). Perishable foods and products go to waste for a variety of reasons, and it isn't necessarily due to a breakdown in the shipping and handling process. However, enhancing sustainability and reducing waste must begin somewhere – and it makes sense to start as near to the source as feasible. IIoT sensors allow for continuous monitoring of perishable food goods in the warehouse. Smart packaging with integrated sensors, for example, may monitor moisture, temperature, and freshness indications. This real-time data allows for the tracking of each product's quality and condition, which enables targeted actions such as prioritizing goods nearing expiration or executing quality inspections.

DHL is a major logistics firm that has integrated IoT and sensor technologies into its operations. DHL employs IoT sensors to track packages' location, temperature, and hydration within, allowing the corporation to optimize its supply chain and improve customer satisfaction (Pal & Kant, 2020; Sharma & Yadav, 2021). Walmart is one corporation that has effectively adopted IoT-based monitoring for perishable items. Walmart tracks and monitors the temperature and humidity of its perishable products, including fresh fruit and frozen items, in real time using IoT sensors and devices. They can discover and fix any faults that may jeopardize the quality of the items by regularly monitoring these parameters. This has assisted Walmart in reducing waste, increasing product freshness, and improving overall supply chain efficiency. For businesses dealing with perishable commodities, IoT-based real-time monitoring has brought various benefits. Companies may reduce losses, preserve product quality, and simplify operations by assuring ideal conditions across the supply chain (Sangeetha & Shunmugan, 2021).

9.3 IIOT ENABLED SMART WAREHOUSES: FUTURE OF INVENTORY MANAGEMENT

Customers' dynamic behavior, fierce rivalry, quick technological advancement, and globalization have all created significant problems in maintaining acceptable quality and appropriate inventory levels in warehousing sectors. To handle perishable commodities, warehouse practitioners must monitor and regulate numerous elements such as enzymes, bacteria, temperature, and humidity. These commodities have a limited shelf life and must be stored and managed carefully to avoid deterioration, assure availability, and generate money (Gupta et al., 2020). The ability to provide services to clients or users while being able to survive is crucial to an enterprise's success. A company's major activity in providing goods to its customers is to have sufficient items available at an acceptable price and within a reasonable period.

Many parts of a business are involved in addressing the quandary. Marketing and design departments are first involved, followed by purchasing and, in certain cases, production. Inventory control is a method of ensuring that things are available for sale to users. Inventory enables a company to provide customer service, logistics, or production when demand cannot be fulfilled by acquiring or manufacturing goods (Wild, 2017). Inventory management is the procedure of planning, organizing, and managing the movement of products and commodities from one site to another and inside a business. It is an important role for organizations of all sizes because it ensures that they have adequate inventory on hand to meet customer demand while keeping prices down. Inventory management may be traced back thousands of years, to the period when people first began to sell products. Early inventory management was quite simple, frequently consisting of merely tallying the number of products in store. However, as firms increased in size and complexity, the demand for increasingly sophisticated inventory management systems emerged (Silver, 2008). Every year, billions of dollars of food deteriorates before reaching customers in North America. Temperature-sensitive items are frequently destroyed in the pharmaceutical sector as a result of improper shipping and storage circumstances. Inventory spoiling may have a substantial influence on a company's bottom line and level of customer service. Smart inventory management solutions, on the other hand, may assist in monitoring the location and storage environment of vulnerable material, hence minimizing spoiling and related financial losses (Bharti, 2016).

Radio-frequency identification (RFID) technology allows digital data contained in RFID tags to be sent through radio waves to an RFID reader. RFID tags are either passive or active. Passive tags are powered by an RFID reader, but active tags have their own source of energy, allowing them to continuously transmit the radio signal. Passive RFID tags are appropriate for tracking perishable goods since they are smaller, thinner, more flexible, and significantly less expensive (Jabbar et al., 2018). Thanks to IIoT activities, manufacturers may translate data from sensors and/or RFID tags into important insights about the location, status, and condition of volatile inventory items. We can take sensor readings or RFID reader data, put it through analytics algorithms, and visualize the results. A warehouse professional, for example, can examine a real-time map of perishable item locations and storage conditions.

Along with reporting and visualization features, IIoT systems may be configured to notify users of specific occurrences. If the temperature in a warehouse exceeds a crucial level, an IIoT system may send a warning to a warehouse worker's mobile app, prompting them to modify the cooling system (Mashayekhy et al., 2022). Similarly, warehouse employees can utilize mobile apps to request information from the database of an inventory-tracking system or to manually operate the inventory-tracking system. For example, an inventory specialist can inquire about how many units of a specific stock keeping unit have fewer than 30 days until expiration.

Difficulties are effectively addressed by new technologies such as the IoT, programmable inventory control platforms, and automatic recovery and storage systems (Maheshwari et al., 2021). Data from the IoT may be used to track adherence to food safety requirements. This information can assist in ensuring that food items are safe for ingestion and that the supply chain is ecologically sustainable (Javaid et al., 2021). Systems that are linked together and share data are referred to as integrated systems. Integrated systems in the context of perishable food supply chains might include enterprise resource planning (ERP) systems, warehouse management systems, and IIoT systems. The integration of these technologies can assist to increase inventory data visibility and make it simpler to make educated inventory management decisions. For example, if a food firm has an ERP system as well as a warehouse management system, the two systems may be connected to exchange inventory-level data. The food industry will be able to observe real-time inventory levels in both systems thanks to this connectivity. This data may be utilized to make intelligent inventory management choices, such as whether to order additional supplies or when to shift goods between warehouses (Chande et al., 2005).

By combining temperature sensor data with HVAC (heating, ventilation, and air conditioning) systems, IIoT technologies allow energy optimization in warehouses. This enables real-time monitoring of environmental factors such as temperature and humidity levels, which provides significant insight into perishable products' storage requirements. The HVAC system may adjust its operation to meet the needs of the warehouse, assuring ideal storage conditions and reducing energy waste. This optimization minimizes energy waste and increases efficiency, hence contributing to sustainability goals by reducing the carbon footprint related to warehouse operations. Proactive maintenance is enabled via real-time monitoring and management of the HVAC system, which triggers alerts to solve faults as soon as they arise. IIoT technologies help to save energy in the long run and lower the risk of equipment malfunctions or inadequacies (Fichtinger et al., 2015).

9.4 SMART WAREHOUSES DRIVEN BY IIOT: A KEY THE ENABLER FOR PREDICTIVE ANALYTICS IN THE PERISHABLE FOOD SUPPLY CHAIN

IoT is a network that links mechanical, digital, and computational objects while eliminating the need for human-computer encounters. At Carnegie Mellon University in 1982, the first Internet-connected Coca-Cola vending machine was created to assess the temperature of cold beverages without physical examination. In 1990, John Romkey designed an Internet-controlled toaster that could be managed remotely. Kevin Ashton, Executive Director of Auto-ID Labs at MIT, created the phrase Internet of Things during a 1999 demonstration for Procter & Gamble (Bhogaraju et al., 2021).

IoT analytics, additionally referred to as predictive analytics, forecasts future occurrences by analyzing previous data and patterns. Models that help in forecasting future events are created using statistical and machine learning approaches. This sort of analytics is critical for making educated business choices such as inventory management and demand forecasting. One of the primary benefits of IoT analytics is the capacity to analyze real-time data points, which is enabled through streaming analytics, a technology that analyzes data as it is created (Cerquitelli et al., 2020).

Predictive analytics can help improve the sustainability of perishable food supply networks. Predictive analytics helps optimize many supply chain elements by utilizing historical data, current trends, and advanced algorithms, resulting in decreased waste, higher efficiency, and enhanced overall sustainability. Predictive analytics may reliably estimate future demand for perishable food goods by analyzing past sales data, seasonal trends, and external influences (e.g., weather, and events). Suppliers may plan production and delivery more effectively with improved forecasting, decreasing the risk of overstocking or stockouts, which can result in food waste and monetary losses (Nozari et al., 2021). Nearly one-third of the food produced each year is wasted or abandoned, according to statistics. Every year, an amazing one billion metric tonnes of food is wasted owing to supply chain inefficiencies, which include harvesting, shipping, and storage. Using fruits and vegetables as an example, a startling 492 million tonnes of fresh products were wasted globally in 2011, owing to insufficient food supply chain management.

The befitting example to cite here would be Walmart, one of the world's largest retailers, which collaborated with the technology firm Zest Labs to create the Eden project. The project's purpose was to decrease food waste and enhance the supply chain efficiency for perishable items like fresh vegetables. The Eden system monitors the temperature, humidity, and other environmental factors of fresh fruit along the supply chain, from the farm to the shop shelves, using predictive analytics and IoT sensors. These sensors' data is then analyzed in real-time using powerful algorithms. Walmart can correctly anticipate the rest shelf life of any item in its inventory by exploiting this data. This data allows Walmart's suppliers and distribution centers to prioritize the delivery of food with shorter shelf life, lowering the chance of products rotting before they reach store shelves. Furthermore, the technology assists in identifying any inefficiencies or inconsistencies in the supply chain that might result in waste. For example, if some trucks or warehouses routinely generate products with lower shelf life owing to poor handling or suboptimal circumstances, changes can be made to address these concerns and reduce waste (Radanliev et al., 2019). Smart warehouses may broadcast their location and track product data thanks to the IIoT. Suppliers, manufacturers, and distribution hubs may plan for item shipment, which minimizes processing times and guarantees optimal resource utilization (Kler et al., 2022).

A recent study in *Forbes* stated that sensor technologies, such as IoT-enabled devices, collect granular data on inventory levels and ambient conditions across supply chain processes. This allows businesses to acquire end-to-end supply chain insight from farmers to customers, assisting them in identifying particular environmental concerns in their operations. Predictive analytics may assist organizations in identifying demand patterns and forecasting future trends by utilizing statistical algorithms, machine learning, artificial intelligence, and internal and external data sources (Beasley, 2023). Furthermore, food companies may utilize data-driven insights to optimize water and energy use, as well as manage climate risks including changing weather patterns and natural catastrophes. Businesses may use predictive analytics to minimize overproduction and optimize trucks and cargo shipping routes. Businesses, for example, might spot seasonal tendencies and plan appropriately by analyzing previous demand data. When demand for a product is low, firms might decrease or alter operations to optimize stock levels and avoid waste. Predictive analytics may also be used to optimize trucking and cargo shipping routes in order to maximize fuel economy. Businesses may discover the most effective routes and minimize unexpected delays by analyzing past traffic data and weather trends. This can help them save fuel and increase the sustainability of their operations (Rehman et al., 2019).

Managing stocks only on the basis of vacations or seasonality is no longer adequate. Millions of individuals are increasing their spending on fresh, pleasurable, but perishable and costly commodities like fruits and seafood. Demand for imported salmon (a popular fish) has surged from 74,000 tonnes in 2003 to 283.8 million tonnes in 2013, according to the AQSIQ Statistics Report. The tremendous demand for perishable consumables has hastened the establishment of different profit-making industries that are connected together, such as cultivating, processing, and transporting perishable goods. These processes are linked throughout every stage of the perishable food supply chain. Real-time analytics and predictive modeling are critical for improving order fulfillment, lowering costs, preserving inventory stability, and capitalizing on actual market benefits. Furthermore, data gives important insight into supply and demand, allowing firms to store inventories in critical markets closer to the end customer (Zhang et al., 2017). Organizations gain a complete picture of their position and the quality level of their logistics facilities by measuring all processes.

Predictive analytics gives firms more control over their operations and allows them to implement measures that promote optimal procedures in order to adapt to changing circumstances. With all of the data created in a warehouse on a regular basis, predictive analytics offers immense potential for evaluating current business models and implementing strategic enhancements to deal with any anticipated market shifts (Sgarbossa et al., 2022). Predictive analytics algorithms can detect patterns and trends, allowing you to predict and minimize hazards before they arise. Data analysis may help you better understand your supply chain, identify possible dangers, and take proactive actions to decrease your exposure to such risks. Furthermore, predictive analytics solutions may assist you in identifying key supply chain variables that can be used to analyze performance and identify opportunities for improvement. For example, you may spot trends in delivery delays, inventory levels, or production output by analyzing big data (Jeble et al., 2018).

9.5 IIOT LEAVING ITS FOOTMARKS IN GLOBAL ORGANIZATIONS

The IIoT is an effective technology for increasing the efficiency and sustainability of perishable food supply chains. IIoT may assist businesses to optimize inventory, decrease waste, improve food safety, and minimize environmental impact by gathering real-time data and connecting it with smart systems. A study in IoT analytics states that a lot of worldwide companies are utilizing IIoT in their smart warehouses for perishable food supply chains. Among these companies are:

- Amazon – Amazon is a forerunner in the usage of IIoT in smart warehouses. To track the location and condition of goods, check energy use, and maintain food safety, the corporation employs a range of sensors and gadgets.
- DHL – DHL is another multinational company that employs IIoT in its smart warehouses. Sensors are used by the firm to track the flow of items in real time, monitor temperature and humidity, and optimize inventory levels.
- Walmart – Walmart employs IIoT in its smart warehouses as well. Sensors are used by the corporation to track the position of items, monitor energy use, and improve picking and packaging efficiency.
- Unilever – Unilever is a multinational food and beverage corporation that employs IIoT in its smart warehouses. Sensors are used by the corporation to track product position, monitor temperature and humidity levels, and increase picking and packaging productivity.

9.6 IMPLICATIONS OF THE STUDY

Implication for Academicians: The IIoT-enabled smart warehouse for studying sustainable development practices has far-reaching ramifications for educators across the board. It is a critical stepping stone for future research endeavors, particularly in supply chain management, sustainability, and industrial engineering. Students can profit from the solid foundations established by researching sophisticated tools and implications of IIoT technology in warehouse and supply chain environments. Another key reason for this is that it allows for cross-industry collaboration. The research emphasizes the significance of combining knowledge from engineering, environmental science, data analysis, and commercial disciplines. Such multidisciplinary teamwork is critical for solving the world's complex and expanding environmental concerns. The study's practical findings can greatly contribute to academic instruction, adding to the body of literature and curriculum improvement. Educators now have actual examples and real-world applications to include in course materials, which will improve students' knowledge of how technology and sustainability cross in the arena of supply chain management. Overall, the research contributes to academic growth, interdisciplinary cooperation, methodological innovation, and enhanced teaching in the realm of academia.

Implication for Practitioners: The research of intelligent warehouses provided by IIoT in connection to sustainable development has significant consequences for business experts. First, it emphasizes the significance of IIoT technology adoption in the smart warehouse market. Businesses that use the IIoT benefits can not only promote sustainable development but also improve overall business efficiency. These findings compel enterprises to carefully consider incorporating IIoT technology into their supply chains, opening the path for more sustainable technologies and procedures. Energy reductions, inventory management, and smart waste reduction measures may help businesses save money. This potential to save expenses emphasizes the need to emphasize sustainability to enhance the bottom line and inspire management to engage in sustainable practices to promote profitability. In addition, the study emphasizes the competitive benefits of using an IIoT-enabled smart warehouse. Customers and partners alike often prioritize initiatives that advance sustainability. This goal may lead to increasing market share, corporate growth, and brand repute. As a result, firms are pushed to include sustainability in their business plan, not just as a duty but also as a competitive advantage.

9.7 CONCLUSION

IIoT has changed the way perishable food supply chains are managed and sustained. IIoT-enabled smart warehousing collects real-time data using sophisticated sensors, which may subsequently be connected with smart systems to increase sustainability. Reduced food waste, improved energy efficiency, and food safety are all part of this.

Furthermore, smart warehousing provided by IIoT helps to reduce the environmental effect of the perishable food supply chain. Delivery truck route optimization saves not just transportation expenses but also carbon emissions and energy usage. Businesses may proactively address sustainability goals by embracing IIoT technology and implementing eco-friendly practices in line with global initiatives to mitigate climate change.

Aside from environmental advantages, IIoT-enabled smart warehousing improves food safety and quality assurance. Any deviations from typical circumstances produce real-time notifications, allowing for fast remedial measures. This guarantees that only safe and high-quality items reach customers, resulting in increased customer trust and loyalty. The IIoT has altered smart warehouse operations, making it a critical driver in the development of long-term fragile food supply chains. Smart warehousing powered by IIoT can optimize inventory, minimize waste, and do predictive analytics and real-time monitoring. As a result, it is a leading technology for transforming how organizations handle and transport perishable items in a sustainable manner. Integrating IIoT in the perishable food supply chain is not only technological progress, but also a commitment to the food industry's future being greener, more efficient, and socially responsible.

REFERENCES

Accorsi, R, R Manzini, A Gallo, A Regattieri, C Mora. 2014. "Energy Balance in Sustainable Food Supply Chain Processes." *Making Food Supply Chains Efficient, Responsive and Sustainable* (pp. 1–12), USA. (Paper presented at the *Third International Workshop on Food Supply Chain (WFSC 2014) conference held in San Francisco (CA)* on November 4–7, 2014).

Alfian, G, M Syafrudin, J Rhee. 2017. "Real-Time Monitoring System Using Smartphone-Based Sensors and Nosql Database for Perishable Supply Chain." *Sustainability*. doi:10.3390/su9112073

Aung, MM, YS Chang. 2014. Temperature Management for the Quality Assurance of a Perishable Food Supply Chain. *Food Control* 40: 198–207.

Beasley, K. 2023 https://www.forbes.com/sites/forbestechcouncil/2023/05/22/three-data-driven-technologies-that-are-making-the-supply-chain-more-sustainable/

Bharti, A. 2016. "Maharaja Agrasen Institute of Management Studies, India." *Handbook of Research on Strategic Supply Chain Management in the Retail Industry* (p. 152). IGI Global. doi:10.4018/978-1-4666-9894-9.ch009

Bhogaraju, SD, KVR Kumar, P Anjaiah. 2021. "Advanced Predictive Analytics for Control of Industrial Automation Process." SD Bhogaraju, KVR Kumar, P Anjaiah, JH Shaik *Innovations in the Industrial Internet of Things (IIoT) and Smart Factory*. Accessed September 20. https://www.igi-global.com/chapter/advanced-predictive-analytics-for-control-of-industrial-automation-process/269600

De Boni, A, G Ottomano Palmisano, M De Angelis 2022. "Challenges for a Sustainable Food Supply Chain: A Review on Food Losses and Waste." *Sustainability*. Accessed September 20. https://www.mdpi.com/2071-1050/14/24/16764

Boyes, Hugh, Bil Hallaq, Joe Cunningham, and Tim Watson. 2018. "The Industrial Internet of Things (IIoT): An Analysis Framework." *Computers in Industry* 101 (October): 1–12. doi:10.1016/J.COMPIND.2018.04.015

Buntak, K, M Kovačić, M Mutavdžija 2019. "Internet of Things and Smart Warehouses as the Future of Logistics." *Tehnički Glasnik* 13: 248–53. doi:10.31803/tg-20190215200430

Caldeira, C, V De Laurentiis, ... S Corrado. 2019. "Quantification of Food Waste per Product Group along the Food Supply Chain in the European Union: A Mass Flow Analysis." *Resources, Conservation, and Recycling*. Accessed September 20. https://www.sciencedirect.com/science/article/pii/S0921344919302721

Cerquitelli, T, N Nikolakis, P Bethaz, S Panicucci, F Ventura, E Macii, S Andolina, et al. 2020. "Enabling Predictive Analytics for Smart Manufacturing through an IIoT Platform." Elsevier, 179–84. doi:10.1016/j.ifacol.2020.11.029

Chande, A., S. Dhekane, N. Hemachandra, N. Rangaraj. 2005. "Perishable Inventory Management and Dynamic Pricing Using RFID Technology." *Sadhana – Academy Proceedings in Engineering Sciences* 30 (2–3): 445–62. Doi:10.1007/BF02706255

Chhotaray, P., Behera, B.C., Moharana, B.R., Muduli, K., Sephyrin, F.T.R., 2023. "Enhancement of Manufacturing Sector Performance with the Application of Industrial Internet of Things (IioT)." *Smart Technologies for Improved Performance of Manufacturing Systems and Services* (pp. 1–19). CRC Press.

da Costa, TP, J Gillespie, X Cama-Moncunill, S Ward, J Condell, R Ramanathan, F Murphy. 2022. A Systematic Review of Real-Time Monitoring Technologies and Its Potential Application to Reduce Food Loss and Waste: Key Elements of Food Supply Chains and IoT Technologies. *Sustainability* 15(1): 614.

Data Bridge Market Research. 2023, February 9. Industrial Internet of Things (IOT) Platform Market to Grow at Astonishing Growth of USD 1809.04 Billion by 2030, Size, Share, Growth Rate, Demand & Trends Forecast. *GlobeNewswire News Room*. https://www.globenewswire.com/en/news-release/2023/02/09/2605303/0/en/Industrial-Internet-of-Things-IoT-Platform-Market-to-Grow-at-Astonishing-Growth-of-USD-1809-04-Billion-by-2030-Size-Share-Growth-Rate-Demand-Trends-Forecast.html

Evtodieva, T. E., D. V. Chernova, N. V. Ivanova, J. Wirth. 2020. "The Internet of Things: Possibilities of Application in Intelligent Supply Chain Management." *Advances in Intelligent Systems and Computing* 908: 395–403. Doi:10.1007/978-3-030-11367-4_38

Fichtinger, J, JM Ries, EH Grosse, P Baker. 2015. "Assessing the Environmental Impact of Integrated Inventory and Warehouse Management." *International Journal of Production Economics*. Accessed September 20. https://www.sciencedirect.com/science/article/pii/S0925527315002406?casa_token=u_xTo4s7R8YAAAAA:mNY4rZyKyJkFuHkxGfqKZgbsN5pjgV3k6sjvYJqmw856XINF2xigtHk2aeCaCNfSUTS9w0E-vPo

Galgal, K.N., M Ray, BR Moharana, BC Behera, K Muduli. 2023. "Quality Control in the Era of IoT and Automation in the Context of Developing Nations." *Smart Technologies for Improved Performance of Manufacturing Systems and Services* (pp. 39–50). CRC Press.

Gupta, Mamta, Sunil Tiwari, and Chandra K. Jaggi. 2020. "Retailer's Ordering Policies for Time-Varying Deteriorating Items with Partial Backlogging and Permissible Delay in Payments in a Two-Warehouse Environment." *Annals of Operations Research* 295 (1). 139–61. Doi:10.1007/S10479-020-03673-X

He, Z, V Aggarwal, SY Nof 2018. "Differentiated Service Policy in Smart Warehouse Automation." *International Journal of Production Research*, 56 (22): 6956–70. Doi:10.1080/00207543.2017.1421789

Jabbar, Sohail, Murad Khan, Bhagya Nathali Silva, and Kijun Han. 2018. "A REST-Based Industrial Web of Things' Framework for Smart Warehousing." *Journal of Supercomputing* 74 (9): 4419–33. Doi:10.1007/S11227-016-1937-Y

Javaid, M, A Haleem, RP Singh, S Rab, R Suman. 2021. "Upgrading the Manufacturing Sector via Applications of Industrial Internet of Things (IioT)." *Sensors International*. Accessed September 20. https://www.sciencedirect.com/science/article/pii/S2666351121000504

Jeble, Shirish, Rameshwar Dubey, Stephen J. Childe, Thanos Papadopoulos, David Roubaud, Anand Prakash. 2018. "Impact of Big Data and Predictive Analytics Capability on Supply Chain Sustainability." *International Journal of Logistics Management* 29 (2): 513–38. Doi:10.1108/IJLM-05-2017-0134/FULL/XML

Jouzdani, J, K Govindan. 2021. "On the Sustainable Perishable Food Supply Chain Network Design: A Dairy Products Case to Achieve Sustainable Development Goals." *Journal of Cleaner Production*. Accessed September 20. https://www.sciencedirect.com/science/article/pii/S095965262033105X

Kler, R, R Gangurde, … S Elmirzaev. 2022. "Optimization of Meat and Poultry Farm Inventory Stock Using Data Analytics for Green Supply Chain Network." R Kler, R Gangurde, S Elmirzaev, MS Hossain, NVT Vo, TVT Nguyen, PN Kumar *Discrete Dynamics in Nature and Society*. Accessed September 20. https://www.hindawi.com/journals/ddns/2022/8970549/

Kumar, A, SK Mangla, P Kumar, M Song 2021. "Mitigate Risks in Perishable Food Supply Chains: Learning from COVID-19." *Technological Forecasting*. Accessed September 20. https://www.sciencedirect.com/science/article/pii/S0040162521000755?casa_token=SAZf8H9JlqkAAAAA:rc1iVaQDKCrG8L9egiFld8GR32ECjjkJ6Sku78Mg0EXOqScwoCFKz6_hyKkmj7LrutFCRIsp1Ks

Lee, CKM. 2018. "Development of an Industrial Internet of Things (IIoT) Based Smart Robotic Warehouse Management System." https://aisel.aisnet.org/confirm2018/43/

Maheshwari, Pratik, Sachin Kamble, Ashok Pundir, Amine Belhadi, Nelson Oly Ndubisi, Sunil Tiwari. 2021. "Internet of Things for Perishable Inventory Management Systems: An Application and Managerial Insights for Micro, Small and Medium Enterprises." *Annals of Operations Research*. doi:10.1007/S10479-021-04277-9

McKinsey Research. https://www.mckinsey.com/mgi/our-research/all-research

Mehroof, K 2019. "A Human-Centric Perspective Exploring the Readiness towards Smart Warehousing: The Case of a Large Retail Distribution Warehouse." *International Journal of Information and Management*. Accessed September 20. https://www.sciencedirect.com/science/article/pii/S0268401218306972?casa_token=b8TPJeag4PsAAAAA:3NCsQsiOoo5_4G64wKhdUtNpl2akoO67HADuC112fVLfnnZRfliSfwLdYjLFw5vYA0A4sBI7aV8

Mashayekhy, Y, A Babaei, XM Yuan, A Xue. 2022. "Impact of Internet of Things (IoT) on Inventory Management: A Literature Survey." *Logistics*. Accessed September 20. https://www.mdpi.com/2305-6290/6/2/33

Musavi, MM, A Bozorgi-Amiri. 2017. "A Multi-Objective Sustainable Hub Location-Scheduling Problem for Perishable Food Supply Chain." *Computers & Industrial Engineering*. Accessed September 20. https://www.sciencedirect.com/science/article/pii/S0360835217303418?casa_token=mbgE9xzymKkAAAAA:Y2QxQ7gJF-retebRo5XNCh9NFoYI1kPlcORa5WeZf59qmybfdt-FtXVa0_pTGp5eIjXl3_2lI8I

Nozari, H, M Fallah, H Kazemipoor. 2021. "Big Data Analysis of IoT-Based Supply Chain Management Considering FMCG Industries." Бизнес-информатика. Accessed September 20. https://cyberleninka.ru/article/n/big-data-analysis-of-iot-based-supply-chain-management-considering-fmcg-industries

Pal, Amitangshu, Krishna Kant. 2020. "Smart Sensing, Communication, and Control in Perishable Food Supply Chain." ACM Transactions on Sensor Networks 16 (1). doi:10.1145/3360726

Posey, Brien, Linda Rosencrance, Sharon Shea.2022. "What Is IIoT (Industrial Internet of Things)? | Definition from TechTarget." 2023. Accessed September 20. https://www.techtarget.com/iotagenda/definition/Industrial-Internet-of-Things-IIoT

Radanliev, Petar, David C De Roure, Jason R C, Rafael, Mantilla Montalvo, Peter Burnap. 2019. "Supply Chain Design for the Industrial Internet of Things and the Industry 4.0." doi:10.20944/preprints201903.0123.v1

Rehman, MH, I Yaqoob, K Salah, … M Imran 2019. "The Role of Big Data Analytics in Industrial Internet of Things." *Future Generation*. Accessed September 20. https://www.sciencedirect.com/science/article/pii/S0167739X18313645?casa_token=aNfmg5N63VoAAAAA:FNhN3-b6vygQ6qzjAAgQOzHrXggHQoD14dFe5yB-JxIev0dPpjzVqqZxPuDf-APmulun35EDIhY

Salama, Mahmoud, Ahmed Elkaseer, Mohamed Saied, Hazem Ali, Steffen Scholz. 2019. "Industrial Internet of Things Solution for Real-Time Monitoring of the Additive Manufacturing Process." *Advances in Intelligent Systems and Computing* 852. Springer Verlag, 355–65. doi:10.1007/978-3-319-99981-4_33/COVER

Sangeetha, AS, S Shunmugan. 2021. "Blockchain for IoT Enabled Supply Chain Management-A Systematic Review." *Fourth International Conference on I-SMAC (IoT in Social)*, Accessed September 20. https://ieeexplore.ieee.org/abstract/document/9243371/

Sgarbossa, F, A Romsdal, … OE Oluyisola 2022. "Digitalization in Production and Warehousing in Food Supply Chains." *The Digital Supply*. Accessed September 20. https://www.sciencedirect.com/science/article/pii/B9780323916141000162

Sharma, S, NS Yadav. 2021. "Ensemble-Based Machine Learning Techniques for Attack Detection." *9th International Conference on Reliability, Infocom, 2021*. Accessed September 20. https://ieeexplore.ieee.org/abstract/document/9596152/

Sharma, Ravi, Balázs Villányi. 2022. "Evaluation of Corporate Requirements for Smart Manufacturing Systems Using Predictive Analytics." *Internet of Things* 19 (100554): 100554. https://doi.org/10.1016/j.iot.2022.100554

Silver, Edward A. 2008. "Inventory Management: An Overview, Canadian Publications, Practical Applications and Suggestions for Future Research." *INFOR* 46 (1 Spec. Iss): 15–28. doi:10.3138/INFOR.46.1.15

Soysal, Mehmet, Mustafa Çimen. 2017. "A Simulation Based Restricted Dynamic Programming Approach for the Green Time Dependent Vehicle Routing Problem." *Computers and Operations Research* 88 (December): 297–305. doi:10.1016/J.COR.2017.06.023

Sverko, M, TG Grbac, M Mikuc. 2022. "Scada Systems with Focus on Continuous Manufacturing and Steel Industry: A Survey on Architectures, Standards, Challenges and Industry 5.0." *IEEE Access*. Accessed September 20. https://ieeexplore.ieee.org/abstract/document/9907002/

Tan, SP, LC Ng, N Lyndon, Z Aman, P Kannan. 2023. "A Review on Post-COVID-19 Impacts and Opportunities of Agri-Food Supply Chain in Malaysia." *Peerj*. Accessed September 20. https://peerj.com/articles/15228/

Verdouw, CN, J Wolfert, … AJM Beulens. 2016. "Virtualization of Food Supply Chains with the Internet of Things." *Journal of Food*. Accessed September 20. https://www.sciencedirect.com/science/article/pii/S026087741530056X

Wang, J, H Wang, J He, L Li, M Shen, … X Tan 2015. "Wireless Sensor Network for Real-Time Perishable Food Supply Chain Management." Accessed September 20. https://www.sciencedirect.com/science/article/pii/S0168169914002919?casa_token=cgocH3vHRcAAAAAA:PSpbBeIrtGvKgn000Hx8tkjPvAqj3VctXVhWuB5BJwN-MIjhKdDLibjVmjpv-0_3R0EATjykzJw

Wild, T. 2017. "Best Practice in Inventory Management." https://books.google.com/books?hl=en&lr=&id=5jQ8DwAAQBAJ&oi=fnd&pg=PP1&dq=21.%09Wild,+Tony.+Best+practice+in+inventory+management.+Routledge,+2017.&ots=tPvjubmdgK&sig=V5yRqfTTWk9lkWpVGWG4IIwnskY

Zhang, Yingfeng, Lin Zhao, Cheng Qian. 2017. "Modeling of an IoT-Enabled Supply Chain for Perishable Food with Two-Echelon Supply Hubs." *Industrial Management and Data Systems* 117 (9): 1890–1905. doi:10.1108/IMDS-10-2016-0456/FULL/HTML

10 An Examination of the Impact of Green Marketing Strategies on Consumer Attitudes towards Environmental Sustainability

Granville Embia and Adimuthu Ramasamy
Papua New Guinea University of Technology, Lae, Papua New Guinea

Manidatta Ray
Birla Global University, Bhubaneswar, India

Kamalakanta Muduli
Papua New Guinea University of Technology, Lae, Papua New Guinea

Dillip Kumar Biswal
Krupajal Engineering College, Bhubaneswar, India

10.1 INTRODUCTION

In recent years, there has been a notable disregard for the theme of environmental care and accountability in boardroom discussions and public forums (Sooklal, 2022) This has enabled corporate executives to evade demands for environmental responsibility by shifting blame onto the government and citizens for their negligent attitudes and the rapid depletion of natural resources such as water, forests, and other ecological habitats (Muduli and Barve, 2015; Biswal et al., 2019; Peppoloni and Di Capua, 2022). Consequently, businesses exhibited minimal effort in their marketing endeavours to promote environmental well-being within the specific contexts where they operate. This enabled firms to partake in comprehensive profit-driven

DOI: 10.1201/9781003349877-10

endeavours, disregarding the well-being of their immediate surroundings and resulting in significant and irreversible harm to the environment (Zivkovic, 2022). Consequently, this has led to the depletion of vital resources, greatly affecting the sustenance of the local community. Nevertheless, in the current business landscape, the exhibition of environmental consciousness has become increasingly important. This is evident through various initiatives, including green marketing, support for local environmental organisations, and the implementation of rigorous compliance policies (Nguyen-Viet, 2023). These efforts have not only bolstered public trust but also captured the attention of governmental entities. From the 1990s to the present day, global warming and climate change have emerged as contentious and delicate topics, sparking considerable discussions among governments, companies, and environmental organisations (Muduli and Barve, 2013; Atehmengo, Idika, & Agbede, 2014). Furthermore, various environmental concerns, including air pollution, water and marine pollution, waste pollution, noise pollution, deforestation, resource depletion, and the escalation of sea levels, have garnered significant attention from policymakers and the public (Gerassimidou et al., 2021). This has led to considerable speculation regarding the failure of businesses to reduce their reliance on the natural environment. Consequently, in response to extensive criticism and negative interactions with consumers, businesses have adopted green marketing strategies, including "eco-labelling, "eco-branding", and "environmental marketing", as a means to alleviate pressures and enhance public consciousness regarding the characteristics and attributes of environmentally sustainable brands. According to Delafrooz, Taleghani, and Nouri (2014), there has been a substantial increase in the consumption of green products, but the use of artificial brands has experienced a notable decline. As a result, the United States of America (USA) has witnessed a significant surge in the prominence and visibility of green marketing since the onset of the 21st century. This is evident in the substantial increase in annual sales amounting to US $250 billion (Shabbir et al., 2020). The substantial cash created through green marketing has a notable impact on global politics and the economy, leading to policy modifications and the adoption of environmentally friendly practises (Wang, Cui and Chang, 2023). As a prominent player in both the political and corporate realms, this windfall revenue has considerable influence. The notion of environmental sustainability has gained significant public acceptability because of its alignment with the core objectives of better marketing performance, namely long-term client retention and maximum value creation by firms (Kumar, 2016; Kardos, Gabor, & Cristache, 2019).

In tandem with the concept of sustainability, green marketing has surpassed its traditional counterpart by placing significant emphasis on three-dimensional elements related to environmental protection. These elements include efforts to mitigate pollution, the development and creation of ecologically sustainable products, and the implementation of improved stewardship processes aimed at responsibly acquiring, utilising, and disposing of products with environmental consciousness (Hart, 1995; Al-Adamat, Al-Gasawneh, & Al-Adamat, 2020). Due to this rationale, the concept of green marketing has gained significant prominence in modern corporate spheres, being regarded on par with other key business objectives. Advertising has become a crucial aspect for businesses, requiring significant portions of their annual expenditures to be allocated towards the promotion of products, services, and business concepts that prioritise the preservation of the ecological system (Jan et al., 2023). The focus of customers who prioritise green products has been redirected by the prevailing economic trends, as they play a crucial role in driving the development of the emerging concept known as green marketing. The emergence of these trends has presented global enterprises with a newfound market opportunity (Shabbir et al., 2020).

The study was carried out with the objective of examining the impact of green-marketing strategies in influencing consumer's attitude towards environmental sustainability, in PNG. The municipal councils of various cities in PNG have collaborated in their efforts to mitigate the environmental impact of plastic products, particularly plastic shopping bags and containers. Their approach involved redirecting their strategies towards reducing and ultimately eliminating the usage of such plastic items, with the aim of addressing the pollution generated by citizens and businesses in the area. Being a maritime province, pollution in this region has dual consequences for the environment, manifesting as both land pollution and sea pollution. These forms of pollution have detrimental impacts on the natural habitats of marine and wildlife species (Liu et al., 2023). As a result, this prompted favourable reactions from companies to implement a comprehensive prohibition on environmentally detrimental plastic bags while also offering biodegradable bags and containers (Ahamed et al., 2021). Moreover, enterprises across various industries have devoted their unwavering enthusiasm and allocated substantial time and resources towards the creation and commercialisation of environmentally sustainable products. These include energy-efficient electrical appliances, organic agricultural produce, paints devoid of lead content, recyclable paper materials, detergents free from phosphates, and reusable containers (de Ruyter et al., 2022). In line with the prevailing trend, financial institutions are actively contributing to the preservation of a pollution-free environment for the benefit of the populace (Kumar, Basavaraj and Soundarapandiyan, 2023). This is achieved through the provision of financial resources and promotional efforts. This study aimed to examine the integration of green marketing strategies within businesses and the impact of this integration on information dissemination, awareness creation, consumer education, and the psychological and behavioural elements of consumers' attitudes towards sustainability and ecopreneurship. The integration of green marketing and sustainable development concerns yields favourable outcomes in the form of effective green campaigns (Kardos, Gabor, & Cristache, 2019). The primary objective of the study was to examine the influence of green branding, eco-labels, and product packaging on customer behaviour. Additionally, the study sought to evaluate the validity and credibility of the proposed hypotheses in order to gain a comprehensive understanding of the efficacy of green marketing strategies (Kardos, Gabor, & Cristache, 2019).

In order to gain a complete understanding about the impact of green marketing approaches on the behaviour of consumers of green products, very important questions need to be answered for a deeper and clearer understanding of the issues highlighted earlier. Thus, the focus of this research identifies gaps in existing green marketing literature and the way it influences the behaviour of individual consumers towards the natural environment. Therefore, the research attempts to render suitable solutions to these subsequent questions (Kardos, Gabor, & Cristache, 2019):

(a) Does eco-labelling (EL) influence consumer attitudes towards environmental sustainability?
(b) Do green packaging and branding (GPB) have a significant relationship with consumer attitudes towards environmental sustainability?
(c) Do green products, premium and pricing (GPPP) have a significant relationship with consumer attitudes towards environmental sustainability?
(d) Is there a significant linkage between consumer attitudes towards environmental sustainability and environmental concerns and beliefs (ECB)?

This chapter encompasses eight important themes. Firstly, the introduction examined green marketing and the responses of businesses across the global arena by setting the objective, identifying the problem, and developing the questions that would guide this study; secondly, the review of literature pertaining to the concepts and evolution of green marketing in the 21st century. This was done by individually considering the variables of the study guided by the research questions, statement of the problem that needed to be resolved, setting the objectives needed to be achieved and testing the hypotheses in order to dispel or accept a definite logic stated. Thirdly, the methodology presented the data collection approaches implemented in this study also highlighting the data collection problems encountered; fourthly, the data analysis tools and statistical tests used and the ensuing results; fifthly, the discussion section which explains the relationships and the importance of making sense of the data gathered; sixthly, the managerial applications that the findings may draw in terms of policy formulation and implementation by organisations; and lastly, limitations of the current study, the opportunity identified and the new direction that this study points towards. In addition, the conclusion offers summary of the findings by seeking to reinforce the objectives of the study and providing practical solutions to the questions that guided this study.

10.2 PROBLEM STATEMENT

Since the 1980s, green marketing gained the recognition and prominence it deserved for advocating for proper business practices and environmentally friendly products (Eneizan, Wahab, Zainon, & Obaid, 2016). Countless studies have since been piloted in developing and developed economies with the USA leading the way in green marketing and green behaviour (Cronin et al., 2011; Ottman, 2017; Chamorro, Rubio, & Miranda, 2009; Correia et al., 2023). As these studies prove, green marketing is still in its infancy in some countries, yet at the same time in other countries it has matured. This is due to the fact that most businesses' and consumers' dependency level on the environment is very significant. Although numerous green marketing–focused studies have been conducted in the past across various geographical contexts, there is very little evidence in literature suggesting the green marketing practices and their impact on the buying behaviour of consumers in Papua New Guinea (PNG), specifically Lae City. The urban drift in Lae City, which has experienced a significant economic expansion and a corresponding surge of inhabitants, has led to an abundance

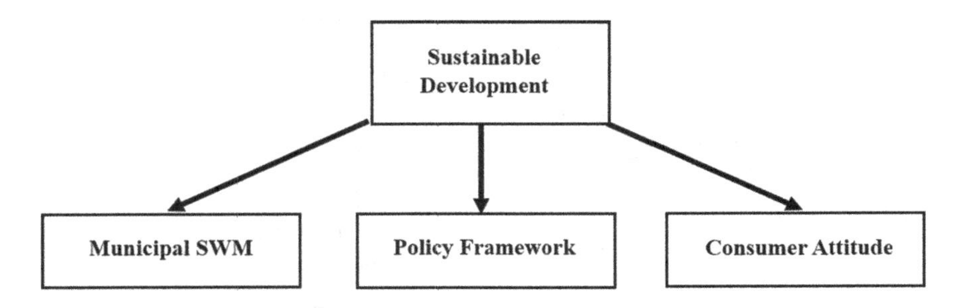

FIGURE 10.1 Factors impacting sustainable development in PNG.

of various resources (e.g. oil, minerals, gas, etc.). However, high development opportunities also pose the risk of substantial waste production. Serious problems with waste reuse or recovery have been caused by subpar municipal solid waste management standards, insufficient funding, and absence of a clear policy framework. The practise has led to contaminated soil and contaminated groundwater, which pose serious risks not only to the health of local residents but also other creatures who are part of the ecosystem. Contextually, it is quintessential to create a mindset among the populace that they may be able to lessen environmental harm. This asks for the creation of sustainable business environment and a green marketing strategy to promote environmental concerns and develop a pro-environmental attitude among the consumers in the region, as depicted in Figure 10.1.

10.3 THEORY BUILDING AND HYPOTHESES DEVELOPMENT

The review of literature on green marketing conjured up numerous researches conducted in the field of marketing and relative literature that warranted quick action from businesses and policymakers by emphasising the significance of green marketing. The concept of green marketing has evolved over time since its first stage in the 1980s where the majority of firms were oblivious to green marketing, making it an unknown concept. The 1990s marked the second stage of this concept where a lot of business were pressured by both consumers and activist groups to show concern for the environment. Marketers realised the disparity that existed between consumers' buying behaviour and their attitudes towards environmentally friendly products. The new millennium marked the third stage of green marketing where technological advancement, tighter government policies, and increased awareness placed much focus on green marketing as a new trend of marketing heading into the future (Punitha & Rasdi, 2013). The green marketing approaches explain how strategies would impact profitability (Eneizan, Wahab, Zainon, & Obaid, 2016). Thus, firms adopted green marketing approaches to provide newer products with superior quality instead of facing the risk of entering new markets and failing (Robins, 2006). Besides, drastic expansion efforts were needed by the firms to ensure that more environmentally friendly ways of maximising value and minimising costs are pursued. A complete overhaul of policy on friendly interactions between humans and nature takes precedence over the conventional consumption mentality to a more

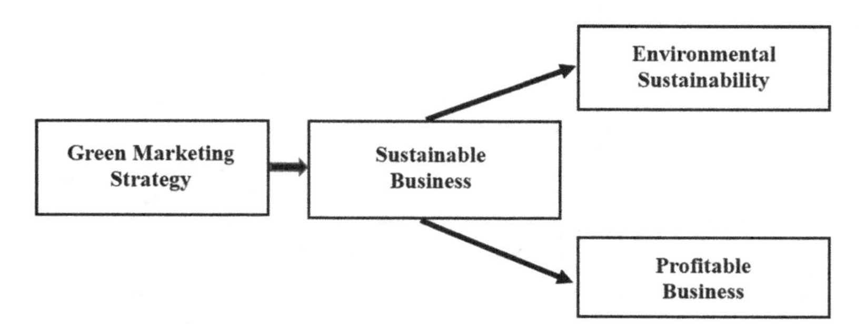

FIGURE 10.2 Sustainable development framework.

ecological protection and welfare analysis mindset for the environment as illustrated in Figure 10.2. Firms, governments and target consumers are working together in redefining value creation and facilitating the transaction through the creation of a biosphere-sustainable practice rather than focusing on traditional consumption marketing (Polonsky, 2011).

10.3.1 Green Marketing

Several researches in the past attempted to explain the vitality of green marketing (Abraham, 2011), none more so than Podvorica & Ukaj (2020), who stated that green marketing included activities and trends such as product modifications, upgrade of productions systems, and improving packaging and labelling with overall advertising strategies of the firm. Welford (2013), pointed that green activities such as identifying, anticipating, and satisfying consumers' requirements were important stages in a firm's green marketing efforts. The task of balancing shifts in the environment as a result of challenges posed by both humans' and firms' demand for safe products (Hasan & Ali, 2015; Juwaheer, Pudaruth, & Noyaux, 2012). To achieve maximum business performance, green marketing tools and strategies were crucial for a firm's long-term sustainability (Papadas, Avlonitis, & Carrigan, 2017).

The American Marketing Association stated that green marketing was marketing environmentally friendly merchandise encompassing these activities such as product alterations, upgrading the systems of production, reforming the packaging, labelling, advertising, and raising awareness about the firm's initiative to do more for the environment (Yazdanifard & Mercy, 2011). The business dictionary defined green marketing as progressive adjustments of consumers' behaviours towards a particular product choice that was affected by a firm's practices and policies that considerably impacted the environment and aroused concerns for the environment and society (Yeng & Yazdanifard, 2015). Thus, green marketing hascontinuously evolved from marketing jargon to a force that strived for sustainability and targeted a segment of consumers that are ecologically sensitive (Drangelico & Vocalelli, 2017). Thus, Polonsky (2008) stated that green marketing was the process of developing strategies aimed at winning consumer segments that exhibit concern for the environment.

Eneizan, Wahab, Zainon, and Obaid (2016) identified important benefits and consequences for firms that adopted green marketing. In terms of benefits, firstly, implementing green practices improved the image and reputation of the firm. Secondly, it increased the firm's revenue through efficient use of resources, increased savings, and optimised use of inputs, resulting in higher profits. Thirdly, implementation of green marketing drastically cut pollution from firms' operations, which enhanced the firm's competitive advantage when compared to non-green businesses through better community goodwill that required them to observe policies and regulations and enabled better practices. On the other hand, there were setbacks associated with green marketing. Firstly, the increased costs of green marketing and investment resulted in distorted marketing efforts. Secondly, low commitment to green marketing forced consumer distrust of the firm. Thirdly, incorrect implementation by a firm led to poor performance affecting its market position. Therefore, continuous refining of green practices yielded better results for the firm. The merits and demerits pertaining to green practices highlighted the complexity and interrelation of environmental problems faced by the world today (Song-Turner, Zeegers, & Courvisanos, 2012).

Now, green consumers were very sensitive and selective about the products that they deemed dangerous to the ecological system and society (Podvorica & Ukaj, 2020). Yet, the agenda of sustainability remained a concern for both business and individuals (Papadopoulos, Karagouni, Trigkas, & Platogianni, 2010). As consumers became aware of environmental problems, consequently their behaviour changed (Cleveland, Kalamas, & Laroche, 2005) by creating an opportunity for green marketing and environmentally sustainable products (Shabbir et al., 2020). However, the gap between favourable attitudes towards environmental concerns and green behaviour still persist (Farzin, Yousefi, Amieheidari, & Noruzi, 2020; Ferraz, Romero, Laroche, & Veloso, 2017). As Shabbir et al. (2020) stated, demonstrating environmentally friendly behaviour significantly impacted on all aspects of the firm. Moreover, products depicting eco-labelling greatly influenced buying behaviour of consumers, in spite of the existing confusion about green products in the mind of the consumers since the 1980s (D'Souza, Taghian, & Lamb, 2006). At the time, consumers of green products endeavoured to maintain a cleaner environment in various ways (Shabbir et al., 2020), despite suspicions about eco-labelling on products (Kardos, Gabor, & Cristache, 2019; Bhaskaran, 2006). Yet more recent studies (Mishra, Jain, & Motiani, 2017; Chen, Hung, Wang, & Huang, 2017; Yang & Zhao, 2019) pointed out that eco-labelling and branding dictated consumer buying efforts. Therefore, green products and their relative pricing strategies influenced consumer buying efforts (Mishra, Jain, & Motiani, 2017), although their beliefs were beginning to rise over time (Ottman, 2017). Thus, information accessibility of green brands must be disseminated in an effective manner that is understood by different segments of green consumers (Mazur, 2014). For an in-depth understanding of green marketing approaches and their impacts on consumer buying behaviour, let us consider the green marketing approaches that impact on consumers (Shabbir et al., 2020).

10.3.2 Perceptions of Environmental Sustainability among Consumers

Past studies revealed three key definitions of green marketing (Eneizan, Wahab, Zainon, & Obaid, 2016). Firstly, green marketing targeted environment-conscious consumers by promoting and producing products that are harmless to the environment. In line with this definition, Banyte, Brazioniene, & Gadeikiene (2010) defined green marketing as segmenting the market to identify and target green consumers by channelling all marketing efforts towards exceeding their expectations and demands. Secondly, Needle (2010) defined green marketing in terms of the triple bottom line objectives based on the conventional marketing mix of the 4Ps (i.e., product, price, promotion, and place). Yet, Violeta & Gheorghe (2009) proposed the six dimensions (5Ps + EE) of green marketing denoted by planning, process, product, promotion, people, and eco-efficiency. Thirdly, Sharma, Lyer, Mehrotra, & Krishna (2010) opined that more effort is needed to connect green consumers to green products using the green marketing mix. Thus, the need for managing the market demand for green products, recycling, and re-manufacturing rise in build-to-order goods and creating competitive advantage headed the environmental objectives of the firm (Sharma, Lyer, Mehrotra, & Krishna, 2010). Given the many aspects of green marketing, several new connotations surfaced to emphasise its importance and raise its profile, such as environmental marketing, environmental marketing management, environmental product differentiation, and eco-friendly labelling (Liu, Kasturiratne & Moizer, 2012).

10.3.3 Eco-Labelling

Labelling a product sends a strong message about the product and its relevance to the consumer. From the marketers' viewpoint, labels of products are designed to position the product in the mind of the consumer different from competing brands. Therefore, eco-labelling is strategically used on environmentally friendly goods as a green marketing approach. It helps in deciding which particular ecological brand to consume and outlines the process through which the product is manufactured (Yeng & Yazdanifard, 2015). Eco-labelling has a positive relationship with consumer's willingness to purchase particular brands. Recognising green labels on products triggers a desire to purchase a product as studies in western nations confirm that consumers are environment-conscious, purchasing eco-labelled products (Cherian & Jacob, 2012). Eco-labels are tools utilised by marketers informing consumers about the effect that a purchasing decision has on the environment (Yeng & Yazdanifard, 2015).

Eco-labelling is significant because it shapes consumer buying behaviour by conveying critical information about consumers' environmental responsibility and quality features of the products (Brécard, 2017) by addressing business and consumers' requirements related to the environmental interactions. The most important role of eco-labels is continuous development of policy guidelines and promotion of environmentally friendly brands and services (Shabbir et al., 2020). On the other hand, confusion still surrounds eco-labelling which results in ambiguity in selecting quality products safe for the environment (Harbaugh, Maxwell, & Roussillon, 2011). Eco-labelling ensures easy recognition of brands and services that have minimal environmental impact throughout their life cycle beginning with the sourcing of raw materials through to the production process and ending with the disposal of the product. Discussions on policy frameworks flourished over time with plans to include eco-labels on all products (Bonroy & Constantatos, 2014) as competition for eco-label products came under strict scrutiny (Fischer & Lyon, 2014). Hence, firms using eco-labelling have been efficient in eliminating substandard products that would compromise the firm's status. Therefore, when handled correctly, eco-labelling becomes a vital approach used by policy architects to create a sustainable and consumable product (Horne, 2009). Bhatia & Jain (2013) reaffirmed that eco-labelling is a handy marketing tool to convince consumers to prioritise green brands. As a result, green marketing accorded firms competitive advantage and a loyal customer base.

10.3.4 Green Packaging and Branding (GPB)

The concept of "green packaging and branding" gained popularity since the 2010s but still remained an unexplored subject in green marketing today. As the calls for better environmental concern grew louder, the exposure of consumers to green packaging and branding resulted in friendly attitudes towards the environment (Swenson & Wells, 2018). Brands changed the attitudes of consumers towards green products through brand differentiation strategies. Commercial success of a firm depended on how much it reflected the green attributes of its products (Hartmann, Ibanez, & Sainz, 2005). Many researchers (Shabbir et al., 2020) argued that green positioning of products was essential to the success of the firms' branding strategies. Other studies (Wustenhagen & Bilharz, 2006) backed green branding by addressing the importance and behaviour of eco-brands. Studies in European countries confirmed that consumers were attracted towards eco-products by demonstrating optimistic attitudes.

10.3.5 Green Products, Premium, and Pricing (GPPP)

As per Swezey & Bird (2001) customers were willing to pay high dollar for green marketing products at a higher than normal price because the products resonated with their belief of utmost ecological responsibility. Green pricing accorded the customers opportunities to invest in renewable energy compared to other products. In some instances, a significant number of consumers of green products have responded favourably to firms' charging premium prices because the quality of the products warranted that much attention (Swezey & Bird, 2001). Furthermore, green regulation pricing strategy tied to excellent environmental products reinforced the firms' competitive advantage (Shabbir et al., 2020) in the business context. The business' profitability (Shabbir et al., 2020) is the direct outcome of the firm's ability to manufacture quality products and its pricing objectives, therefore, setting the appropriate pricing strategy focused on environmental accountability is huge for the firm. So, it is important to consider green pricing and related factors involved in defining the price strategies for brands that are eco-friendly because they influence significant participation of consumers (Shabbir et al., 2020).

10.3.6 Environmental Concerns and Beliefs (ECB)

Environmental concerns and the problems associated with it have affected global consumers (Papadopoulos, Karagouni, Trigkas, & Platogianni, 2010). As a result, consumers have demonstrated concerns about the eco-system and changed their attitudes towards environmental sustainability and safeguarding its welfare (Shabbir et al., 2020). This opened the way for the emergence of green products signified by a vibrant consumer base that patronised eco-products. Active consumers of green products were the reason that environmental protection has made positive headlines worldwide promoting green behaviour that is important for sustainability of the environment (Kardos, Gabor, & Cristache, 2019; Cleveland, Kalamas, & Laroche, 2005; Cleveland & Bartikowski, 2018). The most important behaviour – valuing the environment above anything – else was paramount (Hoyer & MacInnis, 2004). However, concerns were raised (Pickett-Baker & Ozaki, 2008) that many eco-conscious consumers were not demonstrating eco-friendly behaviours that sought to advance the cause of the environment, resulting in a small sample of consumers who performed green activities such as recycling, standing up against pollution and purchasing ecological products (Sloane, 2004).

The development of the subsequent hypotheses aimed at addressing the objectives of the study and assessing the relationship between green marketing approaches and consumers interaction with the environment.

Hypotheses

(a) H1. There is a significant relationship between EL and consumer attitudes towards environmental sustainability (CATES).
(b) H2. There is a significant relationship between the GPB and CATES
(c) H3. There is a significant relationship between the GPPP and CATES
(d) H4. There is a significant relationship between the ECB and CATES.

10.3.7 Theoretical Framework

Constructed from the preceding literature review, the theoretical model was developed as shown in Figure 10.3. The independent variables used for this study comprised of green marketing approaches (Shabbir et al., 2020) such as eco-labelling, green packaging and branding, green product, premium and pricing, and environmental concerns and beliefs. These were used as predictors to measure their impacts on the independent variable, consumer behaviour towards the environment, which

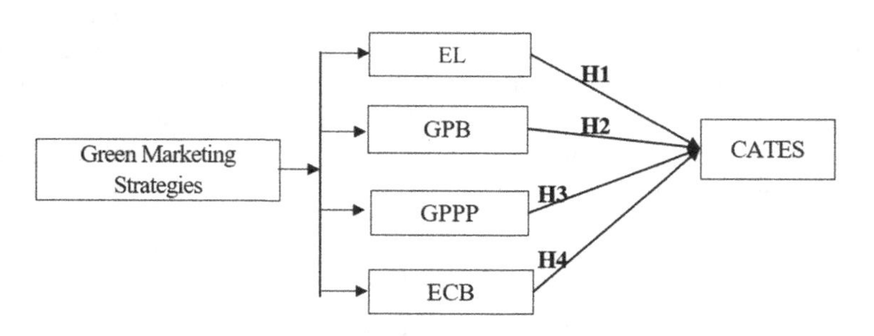

FIGURE 10.3 Theoretical framework.

was being predicted (Karunarathna, Bandara, Silva, & De Mel, 2020). Thus, this framework aimed to comprehend the relationship that was caused by the changes in green marketing approaches and their resulting impact.

10.4 METHODOLOGY

To ensure that the objectives of the study were obtained, a questionnaire was constructed with two sections, Parts A and B (see Appendix A). Part A of the questionnaire was comprised of the four green marketing approaches (Shabbir et al., 2020) included in the questionnaire: eco-labelling (EL), green packaging and branding (GPB), green product, premium, and pricing (GPPP), and environmental concerns and beliefs (ECB). The fifth approach of green marketing included in Part A shed light on the purchasing behaviour of the respondents' consumer belief towards the environment (CBTE). Part B comprised of personal details for demographic profiling of respondents with characteristics such as gender, age, average monthly spending on shopping, occupation, place of origin and province, and signature for verification and authentication purposes. A five-point Likert scale questionnaire was selected for this study to gauge the level of agreeableness to statements under the four green marketing approaches (independent variables) and the consumer's buying behaviour (dependent variable) with 1 = strongly disagree, 2 = disagree, 3 = neutral, 4 = agree and 5 = strongly agree (Dash et al. 2021). The Microsoft Excel version 2013 statistical analysis tool was used to analyse the data gathered. The target population for this study was randomly selected using convenience sampling under the non-probabilistic sampling technique. According to Etikan, Sulaiman, & Alkassim (2016) convenience sampling allowed the researcher to select samples which were conveniently available for the research. Numerous studies by Fernando, Samarasinghe, Kuruppu, and Abeysekera (2017), Michaelidou and Hassan (2017), and Wanninayake and Randiwela (2008) also used this method to survey participants. The participants comprised of consumers from various cities of PNG. A total of 250 questionnaires were distributed, and 150 received in response. After removing the incomplete responses, a total of 110 were found to be usable. The response rate of 44% obtained in this study is considered suitable based on studies of Muduli et al. (2020).

10.5 RESULTS AND ANALYSIS

In this section, the data collected from the participants for this study were analysed to test the hypotheses stated earlier. Using the personal particulars provided, a demographic profile of the respondents using frequency distribution was constructed. Multiple data analysis tools were used to analyse the data and determine whether or not there was a significant relationship between the four green marketing approaches, also known as independent variables, and the predicted variable, which was consumer's behaviour towards the environment. The data analysis tools used for interpretation of the data were Cronbach's alpha test of internal consistency, covariance, and correlation, ANOVA Two factor without replication, and regression analysis (Dash et al., 2021).

10.5.1 Demographic Profile of Respondents

The demographic profiles of the respondents for this study indicated that 63% were females and 37% were males. The age bracket showed that more than half (47%) were 28 years or older and 62 % of the respondents were employed.

10.5.2 Reliability Testing

To measure the reliability of the items used in the survey questionnaire, Cronbach's alpha was employed to ensure that there was internal consistency with all items related to the green marketing approaches variables and their impact on items under the consumer behaviour variable (Shabbir et al., 2020). This test ranges between 0 and 1, with values greater than 0.50 indicating that the questionnaire used to test the hypotheses was more reliable.

Therefore, the internal consistency of the items (individual statements) under the predictor and predicted variables were calculated using the formula $a = 1 - (\textit{mean square errors} \div \textit{mean square rows})$ in 2. Hence, the value of Cronbach's alpha $(a) = 0.80$, shown in Table 10.1, indicates that internal consistency of items was good, hence it was accepted that the items in the questionnaire were consistent with the variables of the study.

10.5.3 Covariance

Here, covariance of the variables was tested to indicate the relationship between the green marketing approaches and their impact on consumer behaviour towards the environment. Therefore, whenever there were changes in the four approaches, "ECB = Environmental concerns & beliefs, EL = Eco-labelling, GPB = Green packaging & branding, GPPP = Green products, premium and pricing would definitely affect CBTE = Consumer behaviour" towards the environment. The resulting figures in Table 10.2 indicate that an increase in the four green marketing tools resulted in an increase in CBTE, which showed a positive covariance. This asserted that changes in the independent variables (ECB, EL, GPB, and GPPP) positively influence CBTE signifying a covariance that was positive.

TABLE 10.1
ANOVA Two-Factors without Replication

Source of Variation	SS	df	MS	F	P-value	F crit
Rows	377.97	74.00	5.11	5.03	0.00	1.29
Columns	425.60	32.00	13.30	13.10	0.00	1.45
Error	2,403.25	2,368.00	1.01			
Total	3,206.82	2,474.00				
Cronbach's Alpha (a)	0.80					

TABLE 10.2
Covariance of the Four Approaches of Green Marketing and Consumer Behaviour

	ECB	EL	GPB	GPPP	CBTE
ECB	0.2506				
EL	0.1197	0.5021			
GPB	0.1299	0.1529	0.3428		
GPPP	0.1126	0.1238	0.1437	0.3891	
CBTE	0.0538	0.0777	0.1485	0.0801	0.2875

10.5.4 Correlation Analysis of Approaches of Green Marketing

In this section, the strength and direction of the relationships between the four green marketing approaches – "*ECB = Environmental concerns & beliefs, EL = Eco-labelling, GPB = Green packaging & branding, and GPPP = Green products, premium and pricing*, and *CBTE = Consumer behaviour towards the Environment*" were tested. Table 10.3 showed that there was a positive relationship between independent and dependent variables, but the more notable relationship was 0.47 between GPB and CBTE. This signified that the better the packaging and branding strategy for green products were, the more consumers were attracted to purchase more green products and act responsibly towards the environment.

10.5.5 Regression Analysis of Green Marketing Approaches

The r-square (r^2) = 0.228 value computed in this research indicates a positive relationship between green marketing approaches and consumers' response to eco-marketing. This indicated that there was a 23% (rounded to the nearest %) variability in *y*, consumer behaviour towards the environment, caused by changes in the independent variables"ECB = Environmental concerns & beliefs, EL = Eco-labelling, GPB = Green packaging & branding, GPPP = Green products, premium and pricing". This signified that the green marketing tools had a significant impact on consumer behaviour.

TABLE 10.3
Correlation Analysis of the Green Marketing Approaches

	ECB	EL	GPB	GPPP	CBTE
ECB	1.00				
EL	0.34	1.00			
GPB	0.44	0.37	1.00		
GPPP	0.36	0.28	0.39	1.00	
CBTE	0.20	0.20	0.47	0.24	1.00

The analysis of variance in Table 10.4 indicated that F-value was 5.182 and significance of 0.001 at $a = 0.05$ (95%). The degree of freedom (df) for regression which was 4 and residual which was 70 where the regression df 4 was the numerator and residual df 70 was the denominator, resulting in significance value of 2.45. The critical value for the F distribution for $a = 0.05$ was 2.45, which is less than the F-value calculated 5.182 at 0.001 level of significance. This indicated that approaches in green marketing approaches can significantly impact the behaviour of consumers towards the environment.

The coefficients of the y – *intercept* (dependent variable) of 1.716 in Table 10.5 indicated a positive linear regression of $y = a + bx$. This signified that "EL = Eco-labelling, GPB = Green packaging & branding, GPPP = Green products, premium, and pricing positively impacts consumer but ECB = Environmental concerns & beliefs" had a negative impact.

10.5.6 Testing of the Hypotheses

The theoretical model depicted the relationship of the green marketing approaches and their impacts on consumers' attitudes and interactions with the environment. The impact of the independent variables (predictors) "(ECB = Environmental concerns & beliefs, EL = Eco-labelling, GPB = Green packaging & branding, and GPPP = Green products, premium and pricing) on the dependent variable (predicted) CBTE = Consumer behaviour towards the environment" was significant. Therefore, various statistical analysis tools were used to test the hypotheses in this study (Karunarathna, Bandara, Silva, & De Mel, 2020).

H1. Eco-labelling (EL) significantly impacts consumer behaviour towards the environment.

H1 was tested using correlation analysis in Table 10.3 to determine whether there was a significant relationship between EL and consumer behaviour towards the environment. The result of EL and CBTE was 0.20 which was the equal lowest relationship coefficient, therefore, there was no significant relationship between EL and CBTE thus H1 was rejected.

H2. Green packaging and branding (GPB) significantly impacts consumer behaviour towards the environment.

TABLE 10.4
Analysis of Variance

	df	SS	MS	F	Significance F
Regression	4	4.927	1.232	5.182	0.001
Residual	70	16.636	0.238		
Total	74	21.563			

H2 was tested using regression analysis in Table 10.5 to determine whether there was a significant relationship between GPB and consumer behaviour towards the environment. The coefficient of GPB and CBTE was 0.413 which was the highest relationship coefficient among the independent variables. Therefore, there was a significant relationship between GPB and CBTE, thus we fail to reject H2.

H3. There is a significant relationship between the green product, premium, and pricing (GPPP) and consumer behaviour towards the environment.

H3 was tested using regression analysis in Table 10.5 to determine whether there was a significant relationship between GPPP and consumer behaviour towards the environment. The coefficients of GPB and CBTE was 0.056 which was the second highest relationship coefficient among the independent variables. Therefore, there was a relationship between GPB and CBTE, thus we fail to reject H3.

H4. Environmental concerns and beliefs (ECB) significantly affects consumer beliefs towards the environment.

H4 was tested using covariance analysis in Table 10.2 to determine the said relationship. The coefficient of GPB and CBTE was 0.054 which was the lowest relationship coefficient among the independent variables. Therefore, there was no relationship between ECB and CBTE, thus we reject H4.

10.6 DISCUSSION

The central aim of this study was to establish whether there was any relationship between the green marketing approaches (ECB = Environmental concerns & beliefs, EL = Eco-labelling, GPB = Green packaging & branding, and GPPP = Green products, premium, and pricing) and their impact on consumer behaviour towards the environment, which is similar to a study conducted in the UAE. Furthermore, this study concentrated on understanding the relationship that green products and green marketing had on consumers' decision-making and ultimate use of green products in an eco-friendly manner. The green marketing approaches helped to position the green brands in the minds of the consumers and differentiated them from firms

TABLE 10.5
Outcome of the Regression

	Coefficients	Standard Error	t-Stat	P-value	Lower 95%	Upper 95%	Lower 95.0%	Upper 95.0%
Intercept	1.716	0.503	3.407	0.001	0.711	2.720	0.711	2.720
ECB	−0.036	0.131	−0.275	0.784	−0.297	0.225	−0.297	0.225
EL	0.024	0.088	0.271	0.788	−0.152	0.199	−0.152	0.199
GPB	0.413	0.115	3.602	0.001	0.184	0.641	0.184	0.641
GPPP	0.056	0.101	0.554	0.581	−0.146	0.259	−0.146	0.259

selling non-green products and also influenced their buying behaviour. Firms were competing for the same resources supplied by nature in attaining the profit goals yet with less environmental responsibility. Thus, consumers now understood the impact that business operations had on the environment and were deviated from conventional marketing mentality of consumption to a more environmentally aware buyer. As a result, firms' revenues, market share, popularity, and other business success attributes declined because social and environment advocacy groups were leading the awareness and promotion efforts about a safe and clean environment. Consequently, green marketing enabled firms to show responsibility and accountability for their actions from sourcing of raw materials to the final sales of the products by making their products eco-friendly. As firms shift to greening, consumers' perception changed and opportunities open for firms positively affecting their profitability. Green marketing did not only benefit the firm and consumers but put the firm in the centre of firms' agenda. It was a win-win strategy (Yeng & Yazdanifard, 2015) that aimed to advance environmental welfare and firm profitability. Customising green marketing to a firm's requirements helped improve the community and the firm's acceptance in society.

This study attempted to understand the relationship between the covariates and the dependent variable, so a covariance analysis was conducted. The findings signify a covariance that was positive. In other words, enriching the green marketing elements, that is, EL, GPB, GPPP, and ECB, resulted in an increase in CBTE. The findings of this study are in line with the study by Ahmed et al. (2023) where the authors conducted the study in the USA. This asserted that changes in the independent variables (ECB, EL, GPB, and GPPP) positively influence CBTE. In our study we also measured the relationship strength between the independent and dependent variables through correlation analysis. As per the findings, there was a notable relationship was between GPB and consumer behaviour towards the environment CBTE. This signifies that the better the packaging and branding strategy for green products were, consumers were attracted to purchase more green products and acted responsibly towards the environment. As per the testing of the hypothesis, a significant relationship was found between the GPB and CATES; this finding is in line with the research by Majeed et al. (2022). Also the relation between the GPPP and CATES was found to be significant and positive which is similar to a study conducted in the UAE by Shabbir et al. (2020).

10.7 MANAGERIAL IMPLICATIONS

Green marketing gained momentum in western countries such as the USA, bringing in yearly revenues of over $250 billion (Shabbir et al., 2020). Global environmental problems such as rising sea levels, environmental pollution, climate change, and rise in greenhouse gas emissions had warranted the adoption and implementation of green marketing. The potential for the green market was vast and the benefits exceeded the costs. Firms could not ignore green marketing but use it as a tool to better the welfare of the society and the firm's objectives. Managers needed to implement it in order to reap its benefits. This study highlighted that EL, ECB, and GPP needed to be improved for consumers to behave in a better green manner. Firstly, the EL on green products did not achieve their purpose because they were not eye catching, lacked sufficient information, were difficult to read, had inaccurate information, and were hard to identify in stores. Secondly, in terms of ECB, consumers depend on the environment for their daily needs so firms that are ignorant towards handling the environment significantly suffer. Last but not least, in terms of GPPP, the portion of price communicates environmental responsibility to consumers, because consumers did not purchase green brands that were applicable to them and upholding transparent production of green products in tandem with environmental diligence was very important. In order to capitalise on the revenue promises that were earmarked for green products, firms and managers alike had to ensure the EL was visible and understandable by customers through pilot testing and marketing research. Furthermore, firms' presence in the community addressed and promoted environmental care as essential. Taking the lead in awareness and promotion of sustainability exercises changed the perception of consumers on firms' practices. In addition, managers advocated for the local application of green products, what portion of price was spent on environmental responsibility, and ensured that products are ecologically manufactured, taking necessary precautions about sensitive environmental problems.

From a practitioner's standpoint, while there are many advantages to using green marketing strategies, they can also present a few challenges, like establishing the authenticity of sustainability claims, dealing with customer scepticism, and navigating complicated regulatory frameworks. Therefore, the success of the strategy depends on having a well-thought-out and authentic green marketing strategy. Directionally, it is crucial for marketers to consider the following aspects related to developing and implementing green marketing tactics.

Ability to get an advantage over competitors – green marketing can give a business an edge in the market, but only if consumers are aware of environmental issues and are inclined to favour goods and companies that show a dedication to sustainability. By developing a position that is based on the sustainability factor, marketers must raise the desired level of awareness among their target audience.

Develop loyalty and advocacy – a green marketing strategy can improve a company's brand image by identifying it with positive environmental and social ideals, which will increase loyalty and advocacy. Customers' faith in the brand may rise as a result of this. However, it is crucial for marketers to be able to forge an emotional connection with their target audience in order to develop brand advocates. In this situation, creating a user profile or persona is essential for providing a better knowledge of consumers and looking for cues that can assist marketers in creating that "emotional bond". A powerful tool in green marketing is effective storytelling. Sharing the brand's sustainability journey, commitment to environmental causes, or the impact of its products can help it connect with customers on a deeper level.

Make your product/brand appealing to consumers – many of them are willing to pay a premium for environmentally friendly products and are more likely to select them than less environmentally friendly alternatives. Green marketing, however, must explicitly define the value proposition, important areas of differentiation, and points of parity of environmentally friendly brands with conventional ones. A "wow factor" can help marketers capitalise on this customer need.

Ensure transparency, accountability and consistency – green marketing demands openness when making sustainability promises. Here, it is crucial for businesses to be able to support their environmental and social claims with reliable evidence, such as statistics, data, and certifications. Additionally, marketers must take responsibility for any mistakes, problems, or complaints that customers have after buying the goods. The adoption of specific crisis/reputation management mechanisms or methods by marketers in this situation is vital, particularly to address any backlash that might arise on social media platforms. The brand communications should also be integrated with the consistency aspect.

Align with long-term goals – green marketing frequently correlates with long-term aims, concentrating on sustainability and corporate responsibility rather than short-term profit maximisation. Companies may also need to construct key performance indicators to measure the success of their green marketing initiatives in terms of long-term growth potential, as well as routinely review progress.

10.8 LIMITATIONS AND FUTURE RESEARCH DIRECTIONS

This study was hindered by the rise in COVID-19 on campus and around the country, which limited visits to firms and population resulting in a small sample collected through convenience sampling method. The lack of financial capacity contributed to sample selection because it would be costly to select a true sample that represents the total population. Providing participants with incentives, such as lunch, coffee, a sandwich, and more, to participate would encourage greater participation, resulting in better outcomes.

There is great opportunity for future research on the theme of green marketing because consumers' and firms' dependence on the environment continue to evolve with time. Thus, a large-scale study can be conducted by evaluating the success of businesses as a result of implement green marketing and the opportunities that it provides for firms with focus on products and their local applications.

10.9 CONCLUSIONS

The green marketing approaches were important for a firm to understand the underlying relationship between these approaches and their impacts on consumers' attitudes towards their surrounding environment. This study analysed the importance of green marketing and noted that more needed to be done to appropriate the maximum efforts of firms and consumers to absorb their impacts and shape perceptions about the environment and the need for conscious interactions.

In addition, firms that integrate the green marketing approaches into their operations create awareness by educating consumers about sustainability, and being ecopreneurs leads environmental responsibility and accountability. Blending green marketing and sustainable eco-campaigns allowed smooth green practices (Kardos, Gabor, &

Cristache, 2019). Moreover, businesses needed to up the ante on EL, ECB, GPB, and GPPP by improving green products through promotions and awareness so that their effects on consumer behaviour towards the environment appeared attractive to consumers and the sustainability of the environment (Kardos, Gabor, & Cristache, 2019).

REFERENCES

Abraham, N. (2011). The apparel aftermarket in India—A case study focusing on reverse logistics. *International Journal of Fashion Marketing Management*, *15*, 211–227.

Ahamed, A., Vallam, P., Iyer, N. S., Veksha, A., Bobacka, J., & Lisak, G. (2021). Life cycle assessment of plastic grocery bags and their alternatives in cities with confined waste management structure: A Singapore case study. *Journal of Cleaner Production*, *278*. https://doi.org/10.1016/j.jclepro.2020.123956

Ahmed, R. R., Streimikiene, D., Qadir, H., & Streimikis, J. (2023). Effect of green marketing mix, green customer value, and attitude on green purchase intention: Evidence from the USA. *Environmental Science and Pollution Research*, *30*(5), 11473–11495. https://doi.org/10.1007/s11356-022-22944-7

Al-Adamat, A., Al-Gasawneh, J., & Al-Adamat, O. (2020). The impact of moral intelligence on green purchase intention. *Management Science Letters*, *10*(9), 2063–2070. https://doi.org/10.5267/j.msl.2020.2.005

Atehmengo, L. N., Idika, I. K., & Agbede, R. I. (2014). Climate change/global warming and its impact on parasitology entomology. *The Open Parasitology Journal*, *5*, 1–11.

Banyte, J., Brazioniene, L., & Gadeikiene, A. (2010). Expression of green marketing developing the conception of corporate social responsibility. *Engineering Economics*, *21*(5), 550–560.

Bhaskaran, S. (2006). Incremental innovation and business performance: Small and medium-size food enterprises in a concentrated industry environment. *Journal of Small Business Management*, *44*, 64–80.

Bhatia, M., & Jain, A. (2013). Green Marketing: A study of Consumer perception and preference in India. *Electronic Green Journal*, *11*(36).

Biswal, J. N., Muduli, K., Satapathy, S., & Yadav, D. K. (2019). A TISM based study of SSCM enablers: an Indian coal-fired thermal power plant perspective. *International Journal of System Assurance Engineering and Management*, *10*, 126–141.

Bonroy, O., & Constantatos, C. (2014). On the economics of labels: How their introduction affects the functioning of markets and the welfare of all participants. *American Journal of Agriculture and Economics*, *97*, 239–259.

Brécard, D. (2017). Consumer misperception of eco-labels, green market structure and welfare. *Journal of Regulatory Economics*, *51*, 340–364.

Chamorro, A., Rubio, S., & Miranda, F. J. (2009). Characteristics of research on green marketing. *Business Strategy and the Environment*, *18*(4), 223–239. https://doi.org/10.1002/bse.571

Chen, Y., Hung, S., Wang, T., & Huang, A. (2017). The influence of excessive product packaging on green brand attachment: The mediation roles of green brand attitude and green brand image. *Sustainability*, *9*, 654.

Cherian, J., & Jacob, J. (2012). Green marketing: A study of consumers' attitude towards environment friendly products. *Asian Social Science*, *8*, 117–126.

Cleveland, M., & Bartikowski, B. (2018). Cultural and identity antecedents of market Mavenism: Comparing Chinese at home and abroad. *Journal of Business Research*, *82*, 354–363.

Cleveland, M., Kalamas, M., & Laroche, M. (2005). Shades of green: Linking environmental locus of control and pro-environmental behaviors. *Journal of Consumer Marketing*, *22*, 198–212.

Correia, E., Sousa, S., Viseu, C., & Larguinho, M. (2023). Analysing the Influence of Green Marketing Communication in Consumers' Green Purchase Behaviour. *International Journal of Environmental Research and Public Health*, *20*(2). https://doi.org/10.3390/ijerph20021356

Cronin, J. J., Smith, J. S., Gleim, M. R., Ramirez, E., & Martinez, J. D. (2011). Green marketing strategies: An examination of stakeholders and the opportunities they present. *Journal of the Academy of Marketing Science*, *39*(1), 158–174. https://doi.org/10.1007/s11747-010-0227-0

Dash, M., Shadangi, P. Y., Muduli, K., Luhach, A. K., & Mohamed, A. (2021). Predicting the motivators of telemedicine acceptance in COVID-19 pandemic using multiple regression and ANN approach. *Journal of Statistics and Management Systems*, *24*(2), 319–339.

Delafrooz, N., Taleghani, M., & Nouri, B. (2014). Effect of green marketing on consumer purchase behaviour. *QScience Connect*, *2014*(1), 1–9.

Drangelico, R., & Vocalelli, D. (2017). Green Marketing: An analysis of definitions, strategy steps, and tools through a systematic review of the literature. *Journal of Clean Production*, *165*, 1263–1279.

D'Souza, C., Taghian, M., & Lamb, P. (2006). An empirical study on the influence of environmental labels on consumers. *International Journal of Corporate Communication*, *11*, 162–173.

Eneizan, B. M., Wahab, K. A., Zainon, M. S., & Obaid, T. F. (2016). Prior research on green marketing and green marketing strategy: Critical analysis. *Singapore Journal of Business Economics and Management Studies*, *5*(5), 1–19.

Etikan, I., Sulaiman, A., & Alkassim, R. (2016). Comparison of convenience sampling and purposive sampling. *American Journal of Theoretical and Applied Statistics*, *5*(1), 1–4.

Farzin, A., Yousefi, S., Amieheidari, S., & Noruzi, A. (2020). Effect of green marketing instruments and behavior processes of consumers on purchase and use of e-books. *Webology*, *17*, 202–215.

Fernando, V., Samarasinghe, D., Kuruppu, G., & Abeysekera, N. (2017). "The Impact of Green Attributes on Customer Loyalty of Supermarket Outlets in Sri Lanka". *in Proceedings of the 3rd International Conference, MERCon*. Moratuwa: University of Moratuwa.

Ferraz, S., Romero, C., Laroche, M., & Veloso, A. (2017). Green products: A cross-cultural study of attitude, intention and purchase behavior. *Review of Administration, Mackenzie*, *18*, 12–28.

Fischer, C., & Lyon, T. (2014). Competing environmental labels. *Journal of Economic Managament & Strategies*, *23*, 692–716.

Gerassimidou, S., Martin, O. V., Chapman, S. P., Hahladakis, J. N., & Iacovidou, E. (2021). Development of an integrated sustainability matrix to depict challenges and trade-offs of introducing bio-based plastics in the food packaging value chain. *Journal of Cleaner Production*, *286*. https://doi.org/10.1016/j.jclepro.2020.125378

Harbaugh, R., Maxwell, J., & Roussillon, B. (2011). Label confusion: The Groucho effect of uncertain standards. *Management Science*, *57*, 1512–1527.

Hart, O. (1995). *Firms, Contracts, and Financial Structures*. Claredon Press, Oxford, UK.

Hartmann, P., Ibanez, V., & Sainz, F. (2005). Green branding effects on attitude: Functional versus emotional positioning strategies. *Marketing Intelligence Planning*, *23*, 9–29.

Hasan, Z., & Ali, N. A. (2015). The impact of green marketing strategy on the firms performance in Malaysia. *Procurement of Social Behavioural Science*, *172*, 463–470.

Horne, R. (2009). Limits to labels: The role of eco-labels in the assessment of product sustainability and routes to sustainable consumption. *International Journal of Consumer Studies*, *33*, 175–182.

Hoyer, W., & MacInnis, D. (2004). *Consumer Behavior*, 3rd ed. Boston, MA, USA: Cengage.

Jan, A. A., Lai, F. W., Siddique, J., Zahid, M., & Ali, S. E. A. (2023). A walk of corporate sustainability towards sustainable development: a bibliometric analysis of literature from 2005 to 2021. *Environmental Science and Pollution Research*, *30*(13), 36521–36532. https://doi.org/10.1007/s11356-022-24842-4

Juwaheer, T., Pudaruth, S., & Noyaux, M. (2012). Analysing the impact of green marketing strategies on consumer purchasing patterns in Mauritius. *World Journal of Entreprenuership and Management of Sustainable Development*, *8*, 36–59.

Kardos, M., Gabor, M., & Cristache, N. (2019). Green marketing's roles in sustainability and ecopreneurship: Case study on Green packaging's impact on Romanian young consumers' environmental responsibility. *Sustainability*, *11*(3), 873.

Karunarathna, A., Bandara, V., Silva, A., & De Mel, D. W. (2020). Impact of green marketing mix on consumer's green purhcasing intention with special reference to Sri Lankan supermarkets. *South Asian Journal of Marketing*, *1*(1), 127–153.

Kumar, P. (2016). State of Green marketing Research over 25 years (1990–2014) literature review and classification. *Marketing Intelligence and Planning*, *34*(1), 137–158.

Kumar, T. P., Basavaraj, S., & Soundarapandiyan, K. (2023). Can co-creating in CSR initiatives influence loyal customers? Evidence from the banking industry. *Corporate Social Responsibility and Environmental Management*. https://doi.org/10.1002/csr.2561

Liu, C., Zhang, X., Xu, Y., Xiang, B., Gan, L., & Shu, Y. (2023). Knowledge graph for maritime pollution regulations based on deep learning methods. *Ocean and Coastal Management*, *242*. https://doi.org/10.1016/j.ocecoaman.2023.106679

Liu, S., Kasturiratne, D., & Moizer, J. (2012). A hub-and-spoke model for multi-dimensional integration of green marketing and sustainable supply chain management. *Industrial Marketing Management*, *41*(4), 581–588.

Majeed, M. U., Aslam, S., Murtaza, S. A., Attila, S., & Molnár, E. (2022). Green marketing approaches and their impact on green purchase intentions: Mediating role of green brand image and consumer beliefs towards the environment.

Mazur, A. (2014). How did the fracking controversy emerge in the period 2010–2012? *Public Understanding of Science*, *25*, 207–222.

Michaelidou, N., & Hassan, L. (2017). The role of health consciousness, food safety concern and ethical identity on attitudes and intentions towards organic food. *International Journal of Consumer Studies*, *32*, 163–170.

Mishra, P., Jain, T., & Motiani, M. (2017). Have green, pay More: An empirical investigation of consumer's attitude towards green packaging in an emerging economy. *In India Studies in Business and Economics; Springer: Singapore*, 125–150.

Muduli, K., & Barve, A. (2013). Modelling the behavioural factors of green supply chain management implementation in mining industries in Indian scenario. *Asian Journal of Management Science and Applications*, *1*(1), 26–49.

Muduli, K., & Barve, A. (2015). Analysis of critical activities for GSCM implementation in mining supply chains in India using fuzzy analytical hierarchy process. *International Journal of Business Excellence*, *8*(6), 767–797.

Muduli, K. K., Luthra, S., Kumar Mangla, S., Jabbour, C. J. C., Aich, S., & de Guimaraes, J. C. F. (2020). Environmental management and the "soft side" of organisations: Discovering the most relevant behavioural factors in green supply chains. *Business Strategy and the Environment*, *29*(4), 1647–1665.

Needle, D. (2010). *Business in Context: An Introduction to Business and Its Environment* (5th edition). South-Western Cengage Learning.

Nguyen-Viet, B. (2023). The impact of green marketing mix elements on green customer based brand equity in an emerging market. *Asia-Pacific Journal of Business Administration*, *15*(1), 96–116. https://doi.org/10.1108/APJBA-08-2021-0398

Ottman, J. (2017). *The New Rules of Green Marketing: Strategies, Tools, and Inspiration for Sustainable Branding* (1st ed.). London, UK: Routledge. https://doi.org/10.4324/9781351278683

Papadas, K., Avlonitis, G., & Carrigan, M. (2017). Green marketing orientation: Conceptualization, scale development and validation. *Journal of Business Research*, *80*, 236–246.

Papadopoulos, I., Karagouni, G., Trigkas, M., & Platogianni, E. (2010). Green marketing: The case of Greece in certified and sustainably managed timber products. *EuroMed Journal of Business*, *5*, 166–190.

Peppoloni, S., & Di Capua, G. (2022). Geoethics: Manifesto for an Ethics of Responsibility Towards the Earth. In *Geoethics: Manifesto for an Ethics of Responsibility Towards the Earth.* https://doi.org/10.1007/978-3-030-98044-3

Pickett-Baker, J., & Ozaki, R. (2008). Pro-environmental products: Marketing influence on consumer purchase decision. *Journal of Consumer Marketing*, *25*, 281–293.

Podvorica, G., & Ukaj, F. (2020). The role of consumers' behaviour in applying green marketing: An economic analysis of the non-alcoholic beverages industry in Kosova. *Wroclaw Review of Law Administration and Economics*, *9*, 1–25.

Polonsky, M. (2008). An introduction to green marketing. *Global Environment: Problems and Policies*, *2*(1), 1–10.

Polonsky, M. (2011). Transformative green marketing: Impediments and opportunities. *Journal of Business Research*, *64*(12), 1311–1319.

Punitha, S., & Rasdi, R. (2013). Corporate social responsibility: Adoption of green marketing by hotel industry. *Asian Social Science*, *9*(17), 79.

Robins, F. (2006). The challenge of TBL: A responsibility to whom? *Business and Society Review*, *111*(1), 1–14.

de Ruyter, K., Keeling, D. I., Plangger, K., Montecchi, M., Scott, M. L., & Dahl, D. W. (2022). Reimagining marketing strategy: Driving the debate on grand challenges. *Journal of the Academy of Marketing Science*, *50*(1), 13–21. https://doi.org/10.1007/s11747-021-00806-x

Shabbir, M. S., Sulaiman, M. A. B. A., Al-Kumaim, N. H., Mahmood, A., & Abbas, M. (2020). Green marketing approaches and their impact on consumer behavior towards the environment—a study from the UAE. *Sustainability (Switzerland)*, *12*(21), 1–13. https://doi.org/10.3390/su12218977

Sharma, A., Lyer, G., Mehrotra, A., & Krishna, R. (2010). Sustainability and business to- business marketing: A framework and implications. *Industrial Marketing Management*, *39*, 330–341.

Sloane, A. (2004). Top 10 Functional Food Trends. *Food Technology*, *58*, 28–51.

Song-Turner, H., Zeegers, M., & Courvisanos, J. (2012). Strategies for marketing greenness: A case study of an architectural design firm in China. *World*, *2*(7).

Sooklal, A. (2022). The Indo-Pacific, an emerging paradigm for peace, cooperation, sustainable development and mutual prosperity. *Journal of the Indian Ocean Region*, *18*(3), 273–281. https://doi.org/10.1080/19480881.2023.2172814

Swenson, M., & Wells, W. (2018). Useful Correlates of Pro-Environmental Behavior. In *Social Marketing: Theoretical and Practical Perspectives* Hove, UK: Goldberg, M.E., Fishbein, M., Middlestadt, S.E., Eds.; Psychology Press; pp. 91–109.

Swezey, B., & Bird, L. (2001). *Utility Green Pricing Programs: What Defines Success?* Golden, CO, USA: National Renewable Energy Laboratory.

Violeta, S., & Gheorghe, I. (2009). The green strategy mix – a new marketing approach. *Knowledge Management and Innovation in Advancing Economics – Analysis and Solutions*, *1–3*, 1344–1347.

Wang, J., Cui, M., & Chang, L. (2023). Evaluating economic recovery by measuring the COVID-19 spillover impact on business practices: Evidence from Asian markets intermediaries. *Economic Change and Restructuring*, *56*(3), 1629–1650. https://doi.org/10.1007/s10644-023-09482-z

Wanninayake, W., & Randiwela, P. (2008). Consumer Attractiveness towards Green Products of FMCG Sector: An Empirical Study". *in the Proceedings of Oxford Business and Economics Conference Program*, Oxford, UK: University of Oxford.

Welford, R. (2013). *Hijacking Environmentalism: Corporate Responses to Sustainable Development*. Routledge: London, UK.

Wustenhagen, R., & Bilharz, M. (2006). Green energy market development in Germany: Effective public policy and emerging customer demand. *Energy Policy*, *34*, 1681–1696.

Yang, Y., & Zhao, X. (2019). Exploring the relationship of green packaging design with consumers' green trust, and green brand attachment. *International Journal of Social Behavioural Perspective*, *47*, 1–10.

Yazdanifard, R., & Mercy, I. (2011). The impact of green marketing on customer satisfaction and environmental safety. *International Conference on Computer Communication and Management*, *5*, 637–641.

Yeng, W. F., & Yazdanifard, R. (2015). Green Marketing: A Study of Consumers' Buying Behaviour in Relation to Green Products. *Global Journal of Management and Business Research: E-Marketin*, *15*(5).

Zivkovic, S. (2022). Sustainability Leadership and Boards: A Conceptual Framework. *Proceedings of the European Conference on Management, Leadership and Governance, 2022-Novem*, pp. 456–463. https://doi.org/10.34190/ecmlg.18.1.587

11 A Framework to Evaluate the Influence of Digital Technology and Sustainable Smart Product-Service Systems on Greenhouse Gas Emissions

Yaone Rapitsenyane, Richie Moalosi, Patrick Dichabeng, Keiphe Setlhatlhanyo, and Oanthata Sealetsa
University of Botswana, Gaborone, Botswana

11.1 INTRODUCTION

While the world continues to look to more technological advancements for solutions, it is equally necessary to shift our production and consumption patterns away from greenhouse gases (GHG)-intensive business models towards more sustainable ones. This shift should offer integrated alternatives for technological solutions aligned with sustainable business model objectives. Earlier research on reducing greenhouse emissions in the food service system (Garnett, 2011) shows that technological mitigation had inadequacies in that there were rebound effects, both environmentally and ethically. Circular economy business models such as sustainable product-service systems (PSS) can reduce greenhouse gas emissions and the risk of the rebound effect. Koide et al. (2022) examined global warming impacts, reviewing consumer-oriented PSS on the life cycle impacts. They found that strategies such as upgrading, repair, refurbishing, and pooling demonstrated moderate to high potential for improvements with a reduced risk of the rebound effect. In comparison, reuse and sharing were associated with higher risks of the rebound effect. The sustainability benefit of implementing sustainable PSS is in prioritising strategies with low risks of backfiring but with high improvement potential. A life cycle approach to implementing sustainable PSS allows for the integration of multiple strategies per each life cycle stage in order to gain the climate change mitigation benefits of sustainable PSS.

 DOI: 10.1201/9781003349877-11

Digital technologies have, over the years, transformed business processes and continue to show positive economic impacts. A lesson for digital business services in reducing GHG emissions is demonstrated by Belousova et al. (2022), who examined the environmental, social, and governance disclosures of leading companies. The study shows that amongst several different actions being carried out towards net zero are services that promote energy efficiency. Big data analytics and artificial intelligence (AI) offer decision support, improving efficiency, effectiveness, market penetration, and customer loyalty. In the transport sector, digital technologies in information and communication technology (ICT) applications have shown potential in reducing GHG emissions in passenger transport. A study by Bieser and Höjer (2021) argues a balanced perspective of the decarbonisation potential for passenger transport through ICT improving transport. They, however, propose a conceptual framework for evaluating the connection between ICT usage, passenger transport, and GHG emissions. Available data shows that reductions in GHG emissions realised by ICT solutions in the energy, buildings, travel, and transport sectors can be significant by 2030. A high reduction potential of about eight gigatonnes of CO2e or 12% of the global emissions by 2030 due to ICT solutions is demonstrated in a study by Malmodin and Bergmark (2015).

Redefining PSS through the integration of ICT gave birth to smart product-service systems, where digitisation converges with servitisation (Chowdhury et al., 2018; Zheng et al., 2019; Behera et al. 2023; Chowdhery & Bertoni, 2018), unleashing a new era of competitiveness of the use of smart technologies to combine products and services for value differentiation opportunities. Advancements in ICT, digitisation technologies, and AI have evolved as smart circular systems and Smart PSS. A smart circular system provides economic sustainability for Smart PSS (Michelini et al., 2017; Li et al., 2021). Sustainable Smart PSS considers the sustainability of cyber-physical resources as an integrated strategy for a circular life cycle approach (Li et al., 2021). How sustainable is Smart PSS? The sustainability of Smart PSS can be evaluated based on a life cycle assessment approach (Maliqi et al., 2022; Bonilla-Alicea et al., 2020) and should take into consideration not just resource usage and waste production, but also greenhouse gas emissions. However, the influence of Smart PSS on GHG emissions can be evaluated by looking at the application of the Smart PSS and the factors involved in delivering the application. This has been exemplified by Negash and Sarmiento (2023) in their case study in the healthcare industry. This chapter looks at how digital technology and sustainable smart PSS can be evaluated for reductions in GHG emissions.

11.2 SUSTAINABLE BUSINESS MODELS FOR THE CIRCULAR ECONOMY

The circular economy (CE) is a novel economic paradigm that prolongs the usage of materials and preserves their worth via offerings and astute solutions. This model is based on two supply chains, forward and reverse, and presents opportunities for businesses to supply solutions and services through the reverse cycle. Transformation towards service businesses is considered an essential solution for expediting a circular economy, but it may also result in a rebound effect. Investigating ways to survive in a competitive market via creative business models and avoiding activities that might harm the environment present opportunities to reconsider product lifespans. Existing sustainable models are based on a linear economic model that causes numerous environmental problems (Antikainen & Valkokari, 2016; Moharana et al., 2023; Piso et al., 2023) and are unsustainable. Consequently, the new area of the circular economy offers opportunities for digitisation. Khan et al. (2021) identified CE and sustainable business models (SBMs) as areas of research that focus on Industry 4.0 and sustainable supply chains. According to Schaltegger et al. (2016), SBMs are valuable tools for defining, assessing, overseeing, and conveying to clients and other stakeholders a business's sustainable value proposition, including how it generates and provides this value and how it protects or replenishes environmental, social, and financial capital outside of its organisational borders. In the circular economy, SBMs are a critical aspect of promoting sustainable development. More context is provided by five SBMs discussed as a product-service system, recovering and recycling, cradle-to-cradle, sharing model, and circular supply chain.

11.2.1 Product-Service System (PSS)

A product-service system represents a business model where a company delivers value by integrating products and services, which collectively contribute to attaining a specific result (Rapitsenyane et al., 2014). These combinations are offered at three overarching levels: product-oriented, use-oriented, and results-oriented PSS. In the last two combinations of this model, instead of being the product owner, the consumer becomes a user of the service. In contrast, in the first combination, the customer owns the product and accesses after-sales services associated with the product. Access to physical services might be restricted or constrained to customers depending on their geographic locations. In contrast, customers may be unable to access a completely digital service where the use of digital technologies is still limited. Sustainability opportunities for this model are explored in various industries such as manufacturing (Costa et al., 2018; Kühl et al., 2018), maritime (Andersen et al., 2013; Solem et al., 2022), and the service industry (Xing et al., 2017; Kim et al., 2015; Tukker, 2004).

11.2.2 Recovering and Recycling

These are another SBM that emphasises the re-utilisation of materials and maximises the value of product returns, focusing on recapturing economic value from discarded products or materials through strategies such as reuse, repair, refurbishment, remanufacturing, upcycling, and recycling (Blomsma & Brennan, 2017). The characteristics include product and component durability, ease of disassembly, and closed-loop material flow back to production (EMF, 2015). However, implementing this model can be challenging because it requires maintaining low unit costs compared with traditional material production. Additionally, strict quality requirements must be met by the by-product which can take time to achieve consistently. The limitations include costs, limited scalability, and uncertainty of returns. The availability of by-products also plays a significant role in ensuring a steady supply of materials, particularly in areas with a lower population and demand. Recycling business models requires an in-depth understanding of product design and material sciences, as well as the ability to manage the unique physical and chemical properties of diverse composite materials. This expertise is necessary for the implementation of value-creating processes, including both downcycling and upcycling, as well as for the recovery and reuse of components and base materials (Lüdeke-Freund et al., 2019). This model is explored and implemented in various contexts and industries, including waste management, manufacturing, environmental and sustainability programmes, and circular economy initiatives, and often employ recovering and recycling models of process and recycle materials like plastic (McDonough & Braungart, 2013).

11.2.3 Cradle-to-Cradle

The cradle-to-cradle (C2C) model is an innovative approach to sustainable product and system design that emphasises creating products and materials with the entirety of their life cycle in mind to recycle or return them to nature without causing harm (Braungart & McDonough, 2009). The foundation of the C2C model is the notion that products and materials ought to be created with infinite recycling potential and zero waste production throughout their life cycles. In this model, products are considered nutrients that can be continuously cycled back into the production process, focusing on eco-effective material flows, renewable energy, and design for disassembly (EMF, 2015). The limitations include high transition costs, a lack of supporting infrastructure, and difficulty achieving scale (Geissdoerfer et al., 2017). Other challenges include selecting materials that are safe, sustainable, and recyclable because finding alternative materials or reformulating existing ones may be required. Redesigning products to be fully recyclable and easy to disassemble can be complex and may require changes to the manufacturing processes and supply chains. Adhering to existing regulations and standards while implementing C2C principles can be challenging because regulations may sometimes align with the model's goals (Braungart & McDonough, 2009). The C2C concept is primarily applied in sustainable design and manufacturing with a focus on designing products and materials for recyclability and reusability. The C2C model has been explored and implemented in various sectors and industries, including textiles and fashion (Rathinamoorthy, 2019), product design and manufacturing (Braungart et al., 2007), building and construction, and renewable energy.

11.2.4 Material Exchange and Sharing Platforms

Material exchange and sharing platforms are critical in reducing carbon emissions and improving resource efficiency, thus promoting the exchange and sharing of materials, enabling increased utilisation of assets through shared access, such as peer-to-peer sharing of underutilised resources (Cohen & Muñoz, 2016). The model is built on shared infrastructure and collaborative consumption of products and services (EMF, 2015). The challenges associated with adhering to regulations related to the exchange of materials, product safety, and liability issues can be challenging

(Lüdeke-Freund et al., 2019). Ensuring that material exchange and sharing platforms are economically viable for all participants, including platform operators, is crucial. Establishing revenue models to sustain platform operations can be challenging (Hamari et al., 2016). Material exchange and sharing platforms are explored and implemented in various sectors and industries, including textiles and fashion; material exchange platforms are used by the textile and fashion industry to swap or sell excess fabric and textile materials, reducing textile waste (McDonough & Braungart, 2017); and in the consumer goods sector, sharing platforms allow individuals to share or exchange products such as tools, electronics, and household items, promoting resource efficiency and reducing the need for new purchases (Hamari et al., 2016); sharing economy platforms also facilitate the sharing of accommodations, transportation, and resources.

11.2.5 The Circular Supply Chain

The model replaces traditional materials with environmentally friendly alternatives to minimise environmental impact. The model is centred within closed-loop supply chains, where materials flow back into the production system through using techniques including recycling, remanufacturing, reuse, and repair (Geissdoerfer et al., 2018). It aims to minimise waste and maximise resource efficiency throughout the supply chain. This model integrates practices, such as reverse logistics and remanufacturing, contributing to sustainable resource management (Blomsma & Brennan, 2017; Pagell & Shevchenko, 2014). However, this process can be challenging because companies must ensure market demand for new products, which may be more expensive, and ensure that the quality of the new materials is satisfactory or better than that of traditional ones. The cost of production for new products must also be reasonable for companies to be willing to adopt new models, or they may need to be more receptive to changes that increase production costs (MacArthur et al., 2013). Implementing a circular supply chain requires efficient reverse logistics systems for collecting, transporting, and processing used products and materials. Managing the return of products and materials can be complex and expensive. The circular supply chain model has been explored within supply chains and logistics management across various industries and sectors, focusing on optimising resource flow and minimising waste throughout the supply chain (Blomsma & Brennan, 2017; Pagell & Shevchenko, 2014), and it has been implemented in electronics, automotive, packaging, and fashion (Lüdeke-Freund et al., 2019).

The concept of a smart circular system represents a visionary step forward in the realm of the circular economy. In accordance with Sustainable Development Goal 12, which calls for the promotion of sustainable patterns of production and consumption (Langley et al., 2021), this approach leverages cutting-edge digital technologies such as the blockchain, AI, Internet of Things (IoT), and big data analytics to create a more interconnected and data-driven approach to circular practises. Using such systems, it is possible to monitor, optimize, and enhance resource utilisation, track product life cycles, and promote sustainability in various ways. It is imperative to note that Industry 4.0 overlooked the importance of social fairness and sustainability

(Fraga-Lamas et al., 2021; Chhotaray et al., 2024; Galgal et al., 2024). Although Industry 5.0 has not yet been imagined as a complete industrial revolution, it complements Industry 4.0. The 5.0 paradigm focuses on enhancing the efficiency of smart factories through technology while minimising environmental and social impacts.

11.3 SUSTAINABLE PSS AND CLIMATE CHANGE MITIGATION

Sustainable Product Service Systems represent a transformative approach to business models, focusing on integrating products and services to achieve more significant sustainability benefits. This concept, evolving from traditional business models, emphasises not just economic benefits but also environmental and social values (Sattari et al., 2020). Sustainable PSS aims to mitigate climate change by GHG emissions, reducing resource consumption by extending product life cycles, promoting sharing, and encouraging the efficient use of resources (Sousa-Zomer & Cauchick-Miguel, 2019). The key to these systems lies in decoupling economic growth from environmental degradation, offering a pathway to sustainable development that aligns with the circular economy principles (Hernandez, 2019). The potential of Sustainable PSS to significantly reduce GHG emissions has been highlighted in various studies. For instance, Koide et al. (2022) demonstrated that upgrading, repairing, refurbishing, and pooling in PSS could lead to moderate to high improvements in environmental impacts, including GHG emission reductions. These strategies, when implemented effectively, can mitigate the environmental impacts associated with the consumption and production of products by optimising resource use and extending the lifespan of products (Schneider et al., 2018). The life cycle approach, integral to Sustainable PSS, further ensures that every stage, from production to disposal, is optimised for minimal environmental impact (Muñoz López et al., 2020).

While Sustainable PSS is geared towards reducing GHG emissions, it is crucial to consider the rebound effect, where increases in consumption or other indirect effects offset gains in efficiency. This phenomenon can undermine the environmental benefits of PSS strategies. For example, the increased efficiency in resource use might lead to lower costs, potentially encouraging greater consumption that paradoxically increases overall GHG emissions. Understanding and mitigating these rebound effects is essential for ensuring that Sustainable PSS achieves its intended environmental goals. Research like a bibliometric analysis of the rebound effect in a product-service system underscores the significance of pinpointing potential strategies to mitigate this occurrence (Alfarisi et al., 2023a). Additionally, the significance of prioritising the rebound effect in the context of product-service systems is underscored in other studies (Alfarisi et al., 2022), emphasising the need for a conceptual structure to challenge the rebound effect in these systems (Alfarisi et al., 2023b).

Table 11.1 provides a concise overview of other climate change mitigation strategies related to Sustainable PSS, each accompanied by a brief description and a relevant source.

More context for these sustainable PSS climate change mitigation strategies can be provided by looking at the business model in the era of digital technologies and how it relates to Industry 4.0.

TABLE 11.1
Climate Change Mitigations Related to Sustainable PSS

Mitigation	Description	Source
Resource Efficiency	Resource efficiency in sustainable PSS is crucial for reducing GHG emissions, minimising waste and energy consumption through the entire product life cycle.	Schöggl et al. (2020)
Extended Product Lifespan	Sustainable PSS models such as repair, refurbishment, and upgrading extend product lifespans, reducing GHG emissions and resource consumption.	Cooper (2010)
Product Design for Sustainability	Sustainable product design in PSS focuses on ease of repair, upgrade, and recycling, reducing environmental impact.	Bhamra and Lofthouse (2007)
Reduction in Overconsumption	Sustainable PSS, including access-based models such as sharing and leasing, helps reduce overconsumption and GHG emissions.	Catulli (2012)
Energy Efficiency	Sustainable PSS aims for energy efficiency by using less energy to deliver the same service, which can be achieved through improved designs and renewable energy sources.	Sorrell (2007)
Behavioural Change	Consumer behaviour is critical in sustainable PSS, where sustainable consumption patterns can significantly reduce GHG emissions.	Young et al. (2009)
Innovation and Collaboration	Innovation in technology and business models, along with stakeholder collaboration, is vital for developing effective, sustainable PSS.	Chesbrough (2003)
Reduced Transportation Emissions	Efficient logistics, local production, and digital over physical products in sustainable PSS can lower transportation emissions.	McKinnon (2008)
Circular Economy Facilitation	The circular economy concept in sustainable PSS emphasises recycling and reuse to close the loop of product life cycles, reducing waste and GHG emissions.	Geissdoerfer et al. (2017)

11.4 DIGITAL TECHNOLOGY (SMART PSS) AND SUSTAINABLE PSS

In the Industry 4.0 age, digital technologies are helping manufacturing organisations accelerate their servitisation process by offering integrated products and services more quickly (Pirola et al., 2020). Lerch and Gotsch (2015) argue that the trends of digitisation and servitisation for manufacturing companies have great and mostly untapped promise. This allows manufacturing organisations to create new value and strengthen existing customer ties. Concurrently, research lines on digitalisation and servitisation combine to create a Smart PSS. Manufacturing organisations have been undergoing a transformation process known as "servitisation" for decades. This process involves a shift in strategy towards new business models built on bundles of products and services, or "Product-Service Systems" (PSS). Leading companies defining new business models integrating sustainable PSS with digital technologies. Though these two sectors have been seen as separate and different subjects in the literature, researchers have recently focused on the connections between digitalisation and servitisation (Kohtamäki et al., 2019; Frank et al., 2019).

Product-service systems have evolved into Smart product-service systems, which include cutting-edge digital technology to improve the performance, sustainability, and efficiency of goods and services. Smart PSS represents convergence of digitisation with servitisation, marking a significant evolution in sustainable business models. This concept involves using smart technologies to enhance the integration of products and services, thereby creating more value and differentiation. Smart PSS leverages advancements in ICT, AI, and digitalisation not only to improve operational efficiencies but also to drive sustainability. The work of Zheng et al. (2019) and Chowdhury et al. (2018) underscores the potential of Smart PSS in delivering innovative solutions that are both economically viable and environmentally friendly. Smart PSS utilises digital frameworks, the Internet of Things, cloud technology, and analytics (Kuhlenkötter et al., 2017). It is primarily characterised as a PSS based on interconnected intelligent products and service systems offering new functions (Chowdhury et al., 2018). Lerch and Gotsch (2015) suggest that Smart PSS is an IT-led value co-creation business approach, continually striving to fulfil unique customer requirements in an eco-friendly manner. This system includes various stakeholders as participants, smart infrastructure as its foundation, intelligent, interconnected products as the instruments and mediums, with the e-services they generate being the essential values provided.

Typical applications of Smart PSS include remote diagnostics, predictive maintenance, remote product monitoring, and equipment optimisation based on operational data (Pirola et al., 2020). For example, IoT devices and sensors are frequently integrated into physical items via smart PSS. These sensors provide real-time monitoring and analysis by gathering information on product performance, usage, and environmental factors. AI algorithms and data analytics can be used to process the information gathered by IoT sensors. This facilitates predictive maintenance, personalised recommendations, and the optimisation of product and service delivery. Smart PSS can be used for remote monitoring and control. For example, using linked devices such as smartphones and tablets, users and service providers can remotely control and monitor smart products and services. This lessens the need for physical involvement while increasing convenience and control.

Other benefits of Smart PSS include an enhanced customer experience. For instance, products and services related to smart homes are made to improve people's quality of life by offering comfort, security, and energy efficiency (Lerch & Gotsch, 2015). Smart PSS strive to cut down on waste and resource usage. Their contributions to sustainability and a circular economy come from optimising product use, maintenance, and end-of-life procedures. The integration of digital technology into PSS has the potential to generate novel business models, including outcome-based, pay-per-use, and subscription services. These approaches match sustainability goals with financial incentives. Smart PSS can monitor and reduce the environmental impact of products and services. Reducing energy use, pollution, and trash production are all part of sustainability product-service systems. In summary, Smart PSS is used in many different industries, ranging from energy management systems and smart home appliances to smart transportation systems such as ride-sharing and electric scooters. These examples highlight how digital technology may be used to improve sustainability.

Despite the opportunities that come with Smart PSS, there are also challenges with social and ethical issues that must be considered. For instance, Smart PSS has brought up issues related to cybersecurity, data privacy, and technological waste (Tsunetomo et al., 2022). When developing and implementing these systems, social and ethical issues must be considered. In summary, Smart PSS utilise digital technology to develop inventive and eco-friendly solutions that maximise efficiency, convenience, and functionality while reducing their environmental effect. In the continuous endeavour to move towards a more sustainable and interconnected society, these systems are fundamental.

11.5 THE RELATIONSHIP BETWEEN DIGITAL TECHNOLOGY, SUSTAINABLE SMART PSS, AND GHG EMISSIONS

Evaluating the sustainability of Smart PSS is crucial for understanding their impact on GHG emissions and overall environmental performance. This should be best done during the design phase to ensure that failure has been designed out and environmental, social, and economic requirements are met during the delivery phase (Liu et al., 2020). Life cycle assessment is used to assess the environmental impacts of Smart PSS, considering factors such as resource consumption, energy use, and waste generation. Studies by Maliqi et al. (2022) and Bonilla-Alicea et al. (2020) highlight the importance of this holistic assessment in ensuring that Smart PSS not only deliver on economic and service quality objectives but also contribute positively to environmental sustainability, particularly in reducing GHG emissions. This assessment helps in identifying areas for improvement and in making informed decisions to enhance the sustainability profile of Smart PSS.

Barquet et al. (2016) identify five key sustainability factors for a PSS business model: applying design for the environment, identifying economic value, promoting behaviour change, acting towards social well-being, and innovating at different levels. Emphasis is still on applying strategies on a life cycle approach, ensuring that benefits are deliberately created around postponing the need for a new product. The sustainability results of Smart PSS have been exemplified by Ufuk (2023) in the fashion industry, linking sustainability strategies to Smart PSS applications to achieve the desired results. For example, the self-customisation strategy is meant to reduce the cost of engaging a design manufacturer and the need for travel to the designer, and the manufacturers leverage distributed manufacturing and co-creation tools. This results in unprecedented sustainability benefits of no physical design resources and a designer, no vehicle emissions to the designer and manufacturer, and no wasting of materials as these products will be made in smart factories. When it comes to co-creating value, it is imperative that we understand how relationships between actors in the ecosystem affect sustainability goals. The value creation system's efficiency should also be sensitive to actions between actors to eliminate redundancy and actions with no value-added regarding sustainability, hence the use of smart and connected systems, where human actions would be inefficient with a high margin of error and counterproductive actions.

In every ecosystem, there should be a net zero, sustainability, or GHG goal tied to a result to be achieved through a specific Smart PSS intervention. Effectively, this can only be successful if a whole systems design and whole life cycle approach is adopted from design to product or service delivery and whole supply chain management (Jin et al., 2017; Embia et al., 2023b). Climate change mitigation strategies for sustainable PSS can be applied to Smart PSS to embody the digitisation aspect of PSS and emissions reductions to be able to avoid the unsustainability of digital technologies in manufacturing (see Table 11.2).

11.6 IMPLICATIONS FOR PRACTICE

It is essential to note the vital link between sustainable activities and economic value creation (Porter & Kramer, 2011). Decoupling economic success from resource consumption is yet to be a standard practice in the industry due to a lack of knowledge, legislation in developing economies, less proactiveness by industry champions and leaders, and associated perceptions of sustainability as an expensive undertaking. Business has also migrated from physical platforms to cyber-physical platforms. The rate at which resources are depleting is faster than the rate at which recovery is happening. This mandates the use of smart and circular business models, which, above all, facilitates infinity recycling in all its forms and promotes resource efficiency. Companies must be guided in the transition towards sustainable business models, as progress towards sustainable development has not yet achieved a desirable scale (Baumgartner & Rauter, 2017). A shift towards collaborative networks and ecosystems from a linear to a smart circular culture of production and consumption is needed. A smart and sustainable product service system is a disruptive sustainable business model applicable in various economic sectors.

- Smart and connected ecosystems for design and manufacturing should be created as early as universities to run trials on sustainable smart PSS models to be continuously rolled out to the industry.
- Every material product will ultimately reach the end of life. Can sustainable Smart PSS deal with absolute zero waste? Non-recyclables should be banned to focus on exhausting the production of multiple generations of products from recyclables.
- The fusion of physical products with digital features, connectivity, and services in Smart PSS unlocks new possibilities for sustainable and efficient value creation. However, further work is needed to leverage this potential fully.
- Research has indicated that as compared to traditional linear business models, strategies including upgrading, repairing, refurbishing, and pooling within PSS have the potential to cut greenhouse gas emissions considerably. The practical implementation of these strategies can optimise resource use and extend product life cycles.
- There is a crucial need to carry out a systematic evaluation and coordination of efforts to optimise the benefits of Smart PSS in reducing GHG emissions and avoid any unintended consequences while leveraging digitally enabled sustainability strategies.

TABLE 11.2

Comparison of Sustainable PSS and Smart PSS in the Context of Climate Change Mitigation and the Function of Digital Technologies in Manufacturing Emissions Reduction

Mitigation	Description	Ecosystem Goal (Examples)	Sustainability Result	Smart PSS Application
Resource Efficiency	Resource efficiency in sustainable PSS is crucial for reducing GHG emissions and minimising waste and energy consumption throughout the entire product life cycle.	Infinite recycling and dematerialisation	Sort recyclable materials and reduce the need for virgin material, produce on real-time demand.	AI-powered robotics, 3D printing
Extended Product Lifespan	Sustainable PSS models such as repair, refurbishment, and upgrading extend product lifespans, reducing GHG emissions and resource consumption.	Build repair ecosystems	Prolong product life spans and reduce the throw-away culture	Pay per unit of consumption/pay per use
Product Design for Sustainability	Sustainable product design in PSS focuses on ease of repair, upgrade, and recycling, reducing environmental impact.	Design for sustainability principles	Increases the value of recyclables and promotes the use of new business models	Augmented reality for redesign and rethink
Reduction in Overcon-sumption	Sustainable PSS, including access-based models like sharing and leasing, helps reduce overconsumption and GHG emissions.	Promote access-based models	Prolong product life cycles, reduce waste, maximise product use, and reduce throwing stuff away	Big data analytics, digital design
Energy Efficiency	Sustainable PSS aims for energy efficiency by using less energy to deliver the same service, which can be achieved through improved designs and renewable energy sources.	Zero-waste design	Avoid creating waste during the design and manufacturing processes	3D digital design, big data analytics
Behavioural Change	Consumer behaviour is critical in sustainable PSS, where sustainable consumption patterns can significantly reduce GHG emissions.	Increase responsible consumption	Encourage and reinforce sustainable consumption habits	IoT consumer behaviour analytics, AI, virtual personal assistance
Innovation and Collaboration	Innovation in technology and business models, along with stakeholder collaboration, is vital for developing effective, sustainable PSS.	Foster innovation-friendly environments	Drive progress in sustainable practices and technologies	Online/digital platforms, distributed manufacturing
Reduced Transportation Emissions	Efficient logistics, local production, and digital over physical products in sustainable PSS can lower transportation emissions.	Minimise logistics footprint	Lower GHG emissions from transport activities	Digital receipts, subscription services, distributed manufacturing
Circular Economy Facilitation	The circular economy concept in sustainable PSS emphasises recycling and reuse to close the loop of product life cycles, reducing waste and GHG emissions.	Establish closed-loop systems	Minimise resource input and waste output	Cyber-physical systems, radio frequency identi-fication tags, digital-twin

11.7 CONCLUSION

The design and application of a comprehensive framework for evaluating the influence of digital technology and sustainable smart product service systems (PSS) on greenhouse gas emissions represent a significant step towards addressing the critical challenge of climate change. As the world grapples with the urgent need to reduce emissions, this framework provides a structured and multidimensional approach to assess the potential impact of innovative technologies and sustainable service models. Promoting sustainability in products and services is crucial for societal change (Fraga-Lamas et al., 2021; Embia et al., 2023a). To achieve this, technology enablers such as green IoT and AI are necessary to make daily lives more environmentally friendly. These technologies have emerged as a research area for reducing carbon emissions (Fraga-Lamas et al., 2021). In addition, the smart circular economy is made possible by the integration of IoT, AI, and edge computing technologies. Bressanelli et al. (2018) have provided insights into IoT, big data, and analytics have played a significant role in the advancement of connected electronics. These technologies enable eight key functions, including the improvement of product design, the attraction of target customers, monitoring and tracking of products, providing technical support, optimising product usage through maintenance, upgrading of the product, and enhancing renovation and end-of-life activities.

The framework offers a valuable tool for various stakeholders, including businesses, policymakers, and researchers. It facilitates data-driven decision-making, allowing organizations to identify opportunities to reduce emissions while enhancing economic sustainability. It also guides policymakers in crafting incentives, regulations, and policies that foster the adoption of eco-friendly PSS. In essence, this framework lays the foundation for a future that is both sustainable and resilient, encouraging the merging of innovation, sustainability, and the reduction of emissions. As it integrates case studies and practical examples, this framework will keep evolving, offering insights, solutions, and strategies to address global warming worldwide. By embracing the concepts and methods detailed in this framework, we can jointly aim for a future that is sustainable and has lower carbon emissions, all while enjoying the advantages of digital technology and intelligent product-service systems.

REFERENCES

Alfarisi, S., Mitake, Y., Tsutsui, Y., Wang, H., & Shimomura, Y. (2022). A study of the rebound effect on the product-service system: Why should it be a top priority? *Procedia CIRP*, 109, 257–262. https://doi.org/10.1016/j.procir.2022.05.246

Alfarisi, S., Mitake, Y., Tsutsui, Y., Wang, H., & Shimomura, Y. (2023a). Bibliometric analysis of a product service system's rebound effect: Identification of a potential mitigation strategy. *Systems*, 11(9). https://doi.org/10.3390/systems11090452

Alfarisi, S., Mitake, Y., Tsutsui, Y., Wang, H., & Shimomura, Y. (2023b). A conceptual structure for challenging the rebound effect of the product-service system. *Procedia CIRP*, 119, 462–467. https://doi.org/10.1016/j.procir.2023.01.009

Andersen, J. A. B., McAloone, T. C., & Garcia i Mateu, A. (2013). Industry-specific PSS: A study of opportunities and barriers for maritime suppliers. In *DS 75-4: Proceedings of the 19th International Conference on Engineering Design (ICED13), Design for Harmonies, Vol. 4: Product, Service and Systems Design*, Seoul, Korea, pp. 19–22.

Antikainen, M., & Valkokari, K. (2016). A framework for sustainable circular business model innovation. *Technology Innovation Management Review* 6(7), 5–12. http://timreview.ca/article/1000

Barquet, A. P., Seidel, J., Seliger, G., & Kohl, H. (2016). Sustainability factors for PSS business models. *Procedia CIRP*, 47, 436–441.

Baumgartner, R.J., Rauter, R., (2017). Strategic perspectives of corporate sustainability management to develop a sustainable organisation. *Journal of Clean Production*, 140, 81e92.

Behera, B. C., Moharana, B. R., Rout, M., & Debnath, K. (2023). Application of Machine Learning in the Machining Processes: Future Perspective Towards Industry 4.0. *Intelligent Manufacturing Management Systems: Operational Applications of Evolutionary Digital Technologies in Mechanical and Industrial Engineering*, 141–156.

Belousova, V., Bondarenko, O., Chichkanov, N., Lebedev, D., & Miles, I. (2022). Coping with greenhouse gas emissions: Insights from digital business services. *Energies*, 15(8), 2745.

Bhamra, T., & Lofthouse, V. (2007). *Design for sustainability: A practical approach.* UK: Gower.

Bieser, J. C., & Höjer, M. (2021). A framework for assessing impacts of information and communication technology on passenger transport and greenhouse gas emissions. In *Environmental informatics* (pp. 235–253). Cham: Springer International Publishing.

Blomsma, F. and Brennan, G., 2017. The emergence of circular economy: A new framing around prolonging resource productivity. *Journal of Industrial Ecology*, 21(3), 603–614.

Bonilla-Alicea, R. J., Watson, B. C., Shen, Z., Tamayo, L., & Telenko, C. (2020). Life cycle assessment to quantify the impact of technology improvements in bike-sharing systems. *Journal of Industrial Ecology*, 24(1), 138–148.

Braungart, M. & W. McDonough. (2009). *Cradle to cradle: Remaking the way we make things*. London: Vintage.

Braungart, M., McDonough, W. and Bollinger, A., (2007). Cradle-to-cradle design: Creating healthy emissions–a strategy for eco-effective product and system design. *Journal of Cleaner Production*, 15(13–14), 1337–1348.

Bressanelli, G., Adrodegari, F., Perona, M., Saccani, N., (2018). Exploring how usage-focused business models enable circular economy through digital technologies. *Sustainability*, 10(3), 639. https://doi.org/10.3390/su10030639

Catulli, M. (2012). What uncertainty? *Journal of Manufacturing Technology Management*, 23(6), 780–793. https://doi.org/10.1108/17410381211253335

Chesbrough, H. (2003). *Open Innovation: The New Imperative for Creating and Profiting from Technology*. Boston: Harvard Business School Press.

Chhotaray, P., Behera, B. C., Moharana, B. R., Muduli, K., & Sephyrin, F. T. R. (2024). Enhancement of manufacturing sector performance with the application of industrial Internet of Things (IIoT). In *Smart Technologies for Improved Performance of Manufacturing Systems and Services* (pp. 1–19). CRC Press.

Chowdhery, S. A. & Bertoni, M. (2018). Data-driven value assessment of packaging solutions, *IFAC-PapersOnLine*, 1(11), 1119–1124, https://doi.org/10.1016/j.ifacol.2018.08.452

Chowdhury, S., Haftor, D., & Pashkevich, N. (2018). Smart product-service systems (Smart PSS) in industrial firms: A literature review. *Procedia CIRP*, 73, 26–31.

Cohen, B., & Muñoz, P. (2016). Sharing cities and sustainable consumption and production: Towards an integrated framework. *Journal of Cleaner Production*, 134, 87–97.

Cooper, T. (2010). *Longer Lasting Products - Alternatives to the Throw-away Society* (1st Edition): Milton Park: Routledge.

Costa, N., Patrício, L., Morelli, N., & Magee, C. L. (2018). Bringing service design to manufacturing companies: Integrating PSS and service design approaches. *Design Studies*, 55, 112–145.

Ellen MacArthur Foundation (EMF). (2015). *Towards a circular economy: Business rationale for an accelerated transition*. Cowes, UK: Ellen MacArthur Foundation.

Embia, G., Mohamed, A., Moharana, B. R., & Muduli, K. (2023ba). Edge Computing-Based Conditional Monitoring. *Intelligent Manufacturing Management Systems: Operational Applications of Evolutionary Digital Technologies in Mechanical and Industrial Engineering*, 249–270.

Embia, G., Moharana, B. R., Mohamed, A., Muduli, K., & Muhammad, N. B. (2023ab). 3D printing pathways for sustainable manufacturing. In *New Horizons for Industry 4.0 in Modern Business* (pp. 253–272). Cham: Springer International Publishing.

Fraga-Lamas, P., Lopes, S.I. and Fernández-Caramés, T.M., (2021). Green IoT and edge AI as key technological enablers for a sustainable digital transition towards a smart circular economy: An industry 5.0 use case. *Sensors*, 21(17), p. 5745.

Frank, A.G., Mendes, G. H. S., Ayala, N. F. & Ghezzi, A. (2019). Servitization and Industry 4.0 convergence in the digital transformation of product firms: A business model innovation perspective, *Technological Forecasting and Social Change*, 141, 341–351, https://doi.org/10.1016/j.techfore.2019.01.014

Galgal, K. N., Ray, M., Moharana, B. R., Behera, B. C., & Muduli, K. (2024). Quality control in the era of IoT and automation in the context of developing nations. In *Smart Technologies for Improved Performance of Manufacturing Systems and Services* (pp. 39–50). CRC Press.

Garnett, T. (2011). Where are the best opportunities for reducing greenhouse gas emissions in the food system (including the food chain)? *Food Policy*, p. 36, S23–S32.

Geissdoerfer, M., Savaget, P., Bocken, N. M. P., & Hultink, E. J. (2017). The circular economy – A new sustainability paradigm? *Journal of Cleaner Production*, 143, 757–768. https://doi.org/10.1016/j.jclepro.2016.12.048

Geissdoerfer, M., Morioka, S. N., de Carvalho, M. M., & Evans, S. (2018). Business models and supply chains for the circular economy. *Journal of Cleaner Production*,190, 712–721.

Hamari, J., Sjöklint, M., & Ukkonen, A. (2016). The sharing economy: Why people participate in collaborative consumption. *Journal of the Association for Information Science and Technology*, 67(9), 2047–2059.

Hernandez, R. J. (2019). Sustainable product-service systems and circular economies. *Sustainability*, 11(19). https://doi.org/10.3390/su11195383

Jin, M., Tang, R., Ji, Y., Liu, F., Gao, L., & Huisingh, D. (2017). Impact of advanced manufacturing on sustainability: An overview of the special volume on advanced manufacturing for sustainability and low fossil carbon emissions. *Journal of Cleaner Production*, pp. 161, 69–74.

Khan, I.S., Ahmad, M.O. and Majava, J., (2021). Industry 4.0 and sustainable development: A systematic mapping of triple bottom line, Circular Economy, and Sustainable Business Models perspectives. *Journal of Cleaner Production*, 297, 126655.

Kim, S., Son, C., Yoon, B., & Park, Y. (2015). Development of an innovation model based on a service-oriented product service system (PSS). *Sustainability*, 7(11), 14427–14449.

Kohtamäki, M., Parida, V., Oghazi, P. Gebauer, H. & Baines, T. (2019). Digital servitisation business models in ecosystems: A theory of the firm, *Journal of Business Research*, 104, 380–392, https://doi.org/10.1016/j.jbusres.2019.06.027

Koide, R., Murakami, S., & Nansai, K. (2022). Prioritising low-risk and high-potential circular economy strategies for decarbonisation: A meta-analysis on consumer-oriented product-service systems. *Renewable and Sustainable Energy Reviews*, 155, 111858.

Kühl, C., Tjahjono, B., Bourlakis, M., & Aktas, E. (2018). Implementation of circular economy principles in PSS operations. *Procedia CIRP*, 73, 124–129.

Kuhlenkötter, B., Wilkens, U., Bender, B., Abramovici, M., Süße, T., Göbel, J., Herzog, M., Hypki, A. & Lenkenhoff, K. (2017). New perspectives for generating smart PSS solutions – Life cycle, methodologies and transformation. *Procedia CIRP*, 64, 217–222, https://doi.org/10.1016/j.procir.2017.03.036

Langley, D.J., van Doorn, J., Ng, I.C., Stieglitz, S., Lazovik, A. & Boonstra, A., (2021). The Internet of Everything: Smart things and their impact on business models. *Journal of Business Research*, 122, 853–863.

Lerch, C. & Gotsch, M. (2015). Digitalized product-service systems in manufacturing firms: A case study analysis. *Technology Management* 58(5):45–52. https://doi.org/10.5437/08956308X5805357

Li, X., Wang, Z., Chen, C. H., & Zheng, P. (2021). A data-driven reversible framework for achieving Sustainable Smart product-service systems. *Journal of Cleaner Production*, 279, 123618.

Liu, L., Song, W., & Han, W. (2020). How sustainable is smart PSS? An integrated evaluation approach based on rough BWM and TODIM. *Advanced Engineering Informatics*, 43, 101042.

Lüdeke-Freund, F., Gold, S. and Bocken, N.M. (2019). A review and typology of circular economy business model patterns. *Journal of Industrial Ecology*, 23(1), 36–61.

MacArthur, Ellen et al. (2013). "Towards the circular economy". *Journal of Industrial Ecology* 2, 23–44.

Maliqi, M., Boucher, X., & Villot, J. (2022, September). Environmental Assessment Methods of Smart PSS: Heating Appliance Case Study. In *IFIP International Conference on Advances in Production Management Systems* (pp. 351–358). Cham: Springer Nature Switzerland.

Malmodin, J., & Bergmark, P. (2015, September). Exploring the effect of ICT solutions on GHG emissions in 2030. In Philippe Fournier-Viger (Ed.), *EnviroInfo and ICT for Sustainability 2015* (pp. 37–46). Atlantis Press.

McDonough, W. & M. Braungart. (2013). *The Upcycle*. New York: North Point.

McDonough, W., & Braungart, M. (2017). *Cradle to cradle: Remaking the way we make things*. New York: North Point Press.

McKinnon, A. (2008). The Potential of Economic Incentives to Reduce CO2 Emissions from Goods Transport Paper presented at the *1st International Transport Forum on Transport and Energy: The Challenge of Climate Change*, Leipzig.

Michelini, G., Moraes, R. N., Cunha, R. N., Costa, J. M., & Ometto, A. R. (2017). From linear to circular economy: PSS conducting the transition. *Procedia CIRP*, 64, 2–6.

Moharana, B. R., Behera, B. C., Syed, S. A., Muduli, K., & Barnwal, S. (2023). Experimental Investigation of Machining NIMONIC 80 Alloy by WEDM Process via Multi-objective Optimisation Techniques: A Sustainable Approach. In *Progress in Sustainable Manufacturing* (pp. 81–96). Singapore: Springer Nature Singapore.

Muñoz López, N., Santolaya Sáenz, J. L., Biedermann, A., & Serrano Tierz, A. (2020). Sustainability assessment of product–service systems using flows between systems approach. *Sustainability*, 12(8). https://doi.org/10.3390/su12083415

Negash, Y. T., & Sarmiento, L. S. C. (2023). Smart product-service systems in the healthcare industry: Intelligent connected products and stakeholder communication drive digital health service adoption. *Heliyon*, 9(2), e13137. https://doi.org/10.1016/j.heliyon.2023.e13137.

Pagell, M. and Shevchenko, A., (2014). Why research in sustainable supply chain management should have no future. *Journal of Supply Chain Management*, 50(1), 44–55.

Pirola, F., Boucher, X., Wiesner, S. & Pezzotta, G. (2020). Digital technologies in product-service systems: A literature review and a research agenda, *Computers in Industry*, 123, https://doi.org/10.1016/j.compind.2020.103301

Piso, K., Mohamed, A., Moharana, B. R., Muduli, K., & Muhammad, N. (2023). Sustainable Manufacturing Practices through Additive Manufacturing: A Case Study on a Can-Making Manufacturer. *Intelligent Manufacturing Management Systems: Operational Applications of Evolutionary Digital Technologies in Mechanical and Industrial Engineering*, 349–375.

Porter, M.E., Kramer, M.R., (2011). Creating shared value. *Harvard Business Review*. 89 (1/2), 62e77. https://hbr.org/2011/01/the-big-idea-creating-shared-value

Rapitsenyane, Y., Bhamra, T. & Trimingham, R. (2014). Competitiveness experiences of Botswana SMEs in the leather industry and their perceptions of sustainability and product service systems, *2014 International Conference on Engineering, Technology, and Innovation (ICE)*, Bergamo, Italy, pp. 1–9, https://doi.org/10.1109/ICE.2014.6871606

Rathinamoorthy, R., (2019). *Circular fashion. In Circular economy in textiles and apparel* (pp. 13–48). Cambridge: Woodhead Publishing.

Sattari, S., Wessman, A., & Borders, L. (2020). Business model innovation for sustainability: An investigation of consumers' willingness to adopt product-service systems. *Journal of Global Scholars of Marketing Science*, 30(3), 274–290. https://doi.org/10.1080/21639159.2020.1766369

Schaltegger, S., Hansen, E., Lüdeke-Freund, F., (2016). Business models for sustainability. A Co-evolutionary analysis of sustainable entrepreneurship, innovation, and transformation. *Organisation & Environment*, 12, 1e26. Seuring, S., Müller, M., 2008. From a literature review.

Schneider, A. F., Matinfar, S., Grua, E. M., Casado-Mansilla, D., & Cordewener, L. (2018). Towards a sustainable business model for smartphones: Combining product-service systems with modularity. Paper presented at the ICT4S2018. *5th International Conference on Information and Communication Technology for Sustainability*.

Schöggl, J.-P., Stumpf, L., & Baumgartner, R. J. (2020). The narrative of sustainability and circular economy - A longitudinal review of two decades of research, *Resources, Conservation and Recycling*, 163, 105073, https://doi.org/10.1016/j.resconrec.2020.105073

Solem, B.A.A., Kohtamäki, M., Parida, V. and Brekke, T. (2022). Untangling service design routines for digital servitization: Empirical insights of smart PSS in maritime industry, *Journal of Manufacturing Technology Management*, 33(4), 717–740. https://doi.org/10.1108/JMTM-10–2020–0429

Sorrell, S. (2007). *The Rebound Effect: An Assessment of the Evidence for Economy-Wide Energy Savings from Improved Energy Efficiency*. https://ukerc.ac.uk/publications/the-rebound-effect-an-assessment-of-the-evidence-for-economy-wide-energy-savings-from-improved-energy-efficiency/

Sousa-Zomer, T. T., & Cauchick-Miguel, P. A. (2019). Exploring business model innovation for sustainability: An investigation of two product-service systems. *Total Quality Management & Business Excellence*, 30(5–6), 594–612. https://doi.org/10.1080/14783363.2017.1317588

Tsunetomo, K., Watanabe, K. & Kishita, Y. (2022). Smart product-service systems design process for socially conscious digitalisation, *Journal of Cleaner Production*, 368, https://doi.org/10.1016/j.jclepro.2022.133172

Tukker, A., (2004). Eight types of product-service system: Eight ways to sustainability? Experiences from SusProNet. *Business Strategy Environment*, 13 (4), 246e260.

Ufuk, G. Ü. R. (2023). Smart product-service systems in fashion industry: A systematic review of sustainability results. *Verimlilik Dergisi*, 57(4), 747–760.

Xing, K., Rapaccini, M., & Visintin, F. (2017). PSS in healthcare: An under-explored field. *Procedia CIRP*, 64, 241–246.

Young, C., Hwang, K., McDonald, S., & Oates, C. (2009). Sustainable Consumption: Green Consumer Behaviour when Purchasing Products. *Sustainable Development*, pp. 18, 20–31. https://doi.org/10.1002/sd.394

Zheng, P., Wang, Z., Chen, C. H., & Khoo, L. P. (2019). A survey of smart product-service systems: Key aspects, challenges, and future perspectives. *Advanced Engineering Informatics*, 42, 100973.

12 Enhancement of Sustainability Practices in Logistics through Adoption of Digital Technologies

Application to PNG Context from Global Perspective

Granville Embia and Kamalakanta Muduli
Papua New Guinea University of Technology, Lae, Papua New Guinea

Tapas Kumar Moharana
National Council of Science Museums, Kolkata, India

Jitendra Narayan Biswal
Einstein Academy of Technology and Management, India

Noorhafiza Muhammad
Universiti Malaysia Perlis, Perlis, Malaysia

Bikash Ranjan Moharana
Papua New Guinea University of Technology, Lae, Papua New Guinea

12.1 INTRODUCTION

In order to be sustainable, one must ensure that social progress, environmental preservation, and economic growth are all balanced while meeting the demands of the present without endangering those of future generations (Wced, 1987). Sustainability ideas are common; however, those ideas may appear differently, in one's view. Firstly,

DOI: 10.1201/9781003349877-12

humans survived in the environment by relying on the environment to meet their basic needs. Also, people should be responsible for their actions, knowing the results of their actions and how it will affect the natural world. Finally, people are responsible to poise the needs of the current generation and include the ones that will come after. Examples of a good practice on sustainability should consider the economic, social and environmental influence of the decision undertaken that may be pursued in many ways by governments, businesses and individuals. Thus, their actions should accommodate the needs of themselves today, including the ones that will come after. This can be achieved meaningfully when they participate in economic, social and environmental activities without totally overusing the resources, and that is very important in decision-making as they take a leap forward in their social advancement in doing businesses (Burton, 1987). Hence, good practice on sustainability is where governments, companies and individuals, when engaging in activities of their nature, consider their decisions and actions in maintaining user-friendly business practices without imposing excessive burdens upon the environment. Information and Communications Technology is often used in countless ways to speed up vehicle efficiency and improve the environment in a sustainable manner (Cullinane, 2014).

With the advent of Industry 4.0, or the Fourth Industrial Revolution, the digitalisation process has intensified, transforming business content and contributing to a more dynamic environment and market structure (Almeida and Correia, 2016). The digitisation process has led to the rapid and increasing growth of manufacturing processes, the improvement of current procedures and practices, the introduction of new technologies, and a huge increase in the volume and scope of industrial production. The main idea behind Industry 4.0 is to apply Internet of Things (IoT) and services using emerging information technologies. This will allow business and engineering processes to be deeply integrated, allowing production to run in a flexible, economical and environmentally friendly manner while consistently maintaining high quality and low costs (Wang et al., 2016). According to the World Commission on Environment and Development, the goal of digitising the industrial sector should be viewed as a revolutionary shift towards more resilient, sustainable and just societies. By 2025, digital technology alone has the potential to significantly cut logistics-related emissions by up to 10% to 12% and contribute to the decarbonisation of the world economy. Therefore, the goal of the sustainable digital logistics ecosystem is to balance sustainability in terms of the economic, social, and environmental dimensions and reflect the interconnections between them. It also aims to rethink digitally based business models and redesign business processes along the supply chain to promote sustainable development (Van Marwyk et al., 2016; Evans, 2017; Kayikci, 2018).

Therefore, the purpose of this investigation is to enforce the use of sustainability practices by respective companies and individuals, including governments, when they engaged in business operations/international trade, or within the jurisdiction of their own countries in that the goods are transferred to them in a more precise and coherent manner. That will be through the use of adapting digital technologies in terms of logistics without damaging the environment.

12.2 SUSTAINABLE DEVELOPMENT PRACTICES

We the current generation consider our future generations through our decisions and actions today when engaging in business activities which involves the use of logistics as a way of moving or transporting goods from a certain place to another location. That occurs either locally or internationally, regardless of the nature of business, causing damage to the environment, apart from the factories' smoke and big machines operating in the large industries to produce those goods or products that we are buying. As a way of reducing the environmental cost/externalities which may cause threat to human beings and the survival of the plants and animals in the environment, we will apply digital technologies on logistics to promote a sustainable practice in our business operation. In this way we are helping to preserve our ecosystem/environment in solving for the current needs of the community/country without causing any danger or threat on the future generations who will use and benefiting from the environment. That is what sustainability/sustainable development practice is all about. Figure 12.1 shows the sustainability involvement in areas such as economic, environment and social, which relies on human action and decision. Thus, in order for our actions to benefit the future generation's welfare, we should include the most vulnerable members of society, which is important for humans in promoting the concept of sustainability. Sustainable development practice is the way in which resources are extracted from the environment to solve for current needs of the community/country without causing any ill effect on the future generation who also will be benefiting from the environment (Ehrhart and Müller, 2010; Wailoni et al., 2022).

12.3 NEED OF SUSTAINABILITY FOR LOGISTICS

Sustainable logistics is an aspect of an enterprise's activities that aims to measure and reduce the environmental impact of logistical activities. Such measures are determined by the possibility of gaining a competitive advantage in the market, which is required by clients. Green logistics, or sustainable logistics, describes the range of steps that supply chain organisations take to reduce their environmental impact, from the extraction of raw materials to the delivery of the finished product to the customer (Larina et al., 2021). Sustainability needs to be increased just like businesses and this will balance the environmental effort. The significance of the logistics sector in

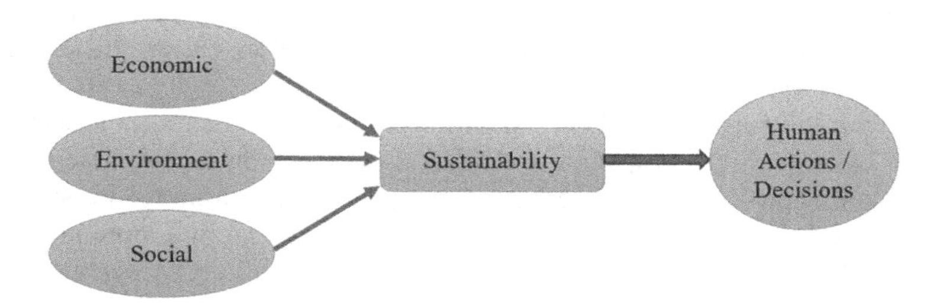

FIGURE 12.1 Sustainability Involvement in Three Areas (Economic, Environment & Social), Which Rely on Human Actions and Decisions.

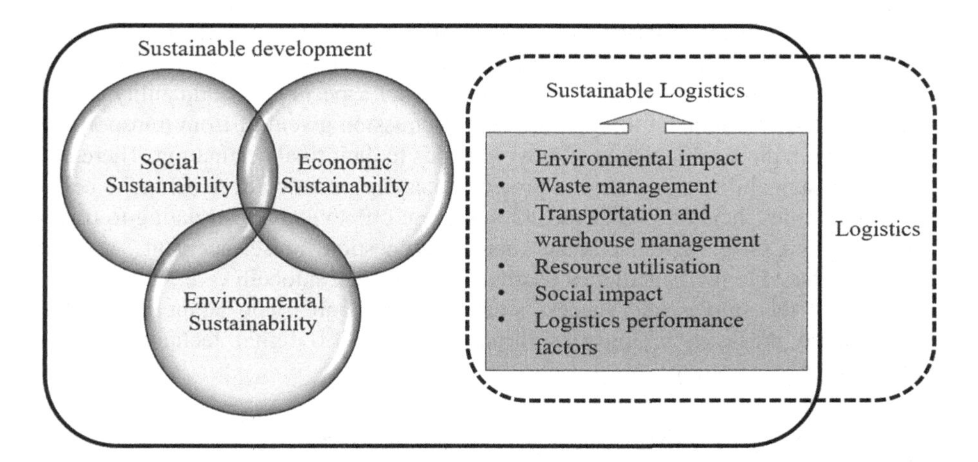

FIGURE 12.2 Concept and Parameters of Sustainable Logistics.

helping a nation succeed in its goals for sustainable development is clear; businesses involved in sustainable logistics will be able to meet a variety of objectives, including reducing environmental pollution, encouraging social advancement, and so on. The following parameters—environmental impact, waste management, transportation and warehouse management, resource utilisation, social impact, and logistics performance factors—form the core values of sustainable logistics (Yontar, 2022). The concept of sustainable logistics and the parameters that affect this are shown in figure 12.2.

Sustainability in logistics is needed simply to reduce emissions in terms of product distribution, which is a concern in the transportation business. Businesses involved in transportation have come to realise that there is an increase in greenhouse gas for every mile travelled and it is totally dangerous to the environment. As per the environmental act concerned, firms have to pay a fine as a form of penalty for harming the environment. However, most firms avoid paying. Hence there is need for sustainable practice to reduce the amount of waste products and improve compliance with government guidelines, customer awareness, and profitability (Rodrigue et al., 2017; Wilson, 2020).

However, a shift to autonomous vehicles and halting of activities is not a requirement at all to an eco-friendly way; instead companies may choose other options which are not costly, with minimal risk associated to it to see improvement in their supply chain's environmental performance. Most rubbish is recycled to avoid paying waste taxes, and is cheaper than viewed by some organisations in the logistics industry who are convinced that it is quite expensive to engage in more environmentally friendly practices. Nevertheless, you have to be mindful of our future generation who have the equal right to live in a harmonious environment just like everybody else today. Therefore, sustainable development practice in logistics is crucial through the application of digital technologies not only to reduce cost and timing and improved efficiency in business operation but for the environmental sustainability as well. The findings of Sandberg and Abrahamsson (2011) concluded that a sustainable competitive advantage is based on a combination of efficient and effective logistics operations and well-functioning, adjusted, in-house developed IT systems. The logistical activity captures both the economic and social aspects of sustainability which will affect the environment through the carbon emission given off from transports as they deal with physical logistics of moving goods to their final destination. There is a need for sustainability in logistics to protect the environment by decreasing the negative externalities including other factors that arose due to activities relating to transport such as CO_2 emission, climate change, congestion, noise, accidents, and air pollution, due to logistics activities (Larina et al., 2021; Centobelli et al., 2020).

Some of the studies and success stories of companies who adapt sustainable development practices in logistics through the use of digital technologies are explained in next section.

12.3.1 AI System 'ORION' Brings Success for UPS

UPS, as a company that engages in distributing goods to regions, should be conscious about their activities as they will contribute to a larger amount of greenhouse gas emission, therefore they surely are required to embark on sustainable development practices by simply adopting digital technology in logistic activities. This is very important for companies such as UPS. In fact, the company had adopted a more sustainable practice since 2012, by developing an AI system named ORION. This is a way forward to them as the system had benefited the company in terms of providing efficacy in transportation, thus promoting route optimization and improving delivery greatly. UPS, through its AI System, ORION, has benefited not only in terms of finance. However, the system has the potential to reduce 100,000 carbon footprint metrics tonnes annually. Also, UPS is saving up to 10 million fuel gallons annually, which resembles reducing approximately 20,000 vehicles on the roads (Destiche, 2013; Horner, 2016).

12.3.2 IKEA and Big Data Analysis to Identify the Best Paths to Success

Businesses may use software tools instead of building hardware which would be very costly to them. Upon using software, offered by public cloud, they are required only to meet the subscription fee when it expires as per their plan.

IKEA made significant progress in its outmoded data method, which was siloed and inaccessible to many departments within the firm. IWAY, who is the supplier code of conduct of IKEA, is another successful implementer of big data analytics. Their aim is to translate the data into useful information to provide an insight on their business operation. Most manufacturing companies are liable to protect the environment by being responsible for their waste product. Hence, IKEA has been a strong advocate for environmental sustainability for 20 years. IKEA is improving its services as per its previous involvements to a newer version, IWAY six, of IKEA's supplier code of conduct. They assess areas on rights of core workers, workplace safety, employee work-life balance, water and waste management of potential suppliers, and prevention of child labour (IKEA Global; Fröding and Lawrence, 2010; Laurin and Fantazy, 2017; Han, 2022).

Data analytics application is not a new thing to the banking industry, telecom, retail, insurance, e-commerce, and airlines companies as they have been using it. Most businesses in India are at the infant stage of big data-driven supply chain decision-making. Supply chains will not operate well without applying big data analytics and data mining. The operation of most logistics companies relies on data and analytics. Logistics industries are experiencing the impact of digital technologies similar to other industries. Data has been influenced by new technologies allowing companies to improve their data by using it with dynamic applications to promote the operation of a sustainable supply chain. This brings many benefits not only to the company but to the environment as well.

12.4 IMPLEMENTATION PLANNING FOR SUSTAINABLE DEVELOPMENT PRACTICES IN LOGISTICS

Implementing sustainable development practices in logistics will involve reducing the damages in the environment from logistics operations in maintaining or improving their efficiency and effectiveness. Through optimising transportation, reducing package waste, using energy-efficient equipment, incorporating reverse logistic processes, and collaborating with suppliers and customers all plays a role in reducing the environmental impact of the entire supply chain. Monitoring and reporting sustainability performances will improve the process of implementing sustainable development practices in logistics as will training employees in waste reduction, energy efficiency, and responsible transportation. Figure 12.3 explains the various aspects needed to implement effective training of stakeholders for sustainable logistics. In order to implement a successful sustainable practice on logistics, employees' training on energy efficiency, responsible transportation, and how to reduce waste is vital for their awareness.

To optimise transportation, we should consider the routes and modes to reduce fuel consumption, emissions, and transportation costs. Reduce package waste, for example, by using eco-friendly packaging materials and design packaging to reduce waste and increase recyclability. Companies must use energy-efficient equipment to reduce energy consumption and greenhouse gas emissions. This equipment includes

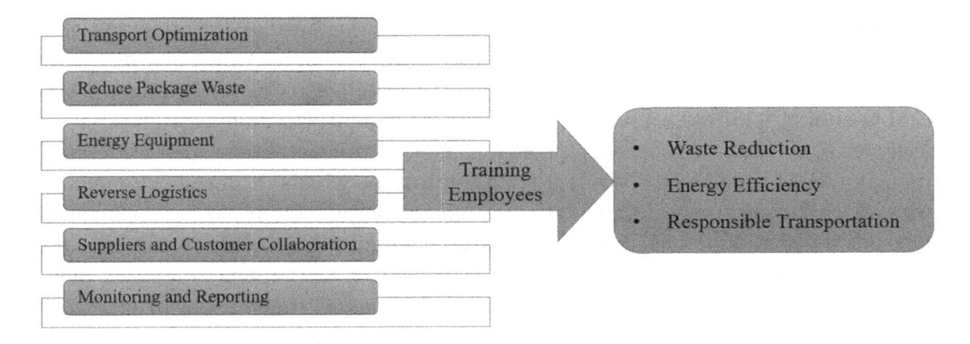

FIGURE 12.3 Process of implementing sustainable practices in logistics.

the use of electric forklifts, LED lighting, and of course, solar panels. All companies are advised to use this equipment. Also, the implementation of the reverse logistics process is encouraged, to be applied to recover and recycle materials and products, reducing waste and supporting the circular economy. Collaborating with suppliers and customers includes sourcing of sustainable produced materials, reducing packaging waste, and optimising transportation routes. Monitoring and reporting sustainability performance promotes the progress of sustainability goals by demonstrating a clear path to improve if need be. In fact, the idea of training employees on sustainable development practices in relation to waste reduction, energy efficiency, and responsible transportation is a prerequisite to a sustainable development practice in logistics by companies. Thus, all these contribute to a better implementation of a sustainable development practice in logistics. Economically, by improving sustainable development practices in logistics, these will improve the company's reputation and potentially reduce costs while maximising operation.

12.5 IMPLEMENTATION OF SUSTAINABLE PRACTICES IN LOGISTICS AT PAPUA NEW GUINEA (PNG)

Sustainable development practices can be implemented in logistics activities in PNG by employing digital technologies such as AI, IoT, and big data analytics. By adapting these technologies towards a portfolio of sustainable practices, this will not only increase performance but help to enhance the functionality of circular supply chains in the operating business.

According to Goundar et al. (2022), AI can help to optimise logistics operation by predicting demand and supply patterns. IoT can help to track goods and monitor environmental conditions during transportation. Big data analytics will be helpful to analyse the large amounts of data generated by logistics activities to identify inefficiencies and opportunities for improvement (Muduli and Barve, 2013; Yang et al., 2021).

12.5.1 Artificial Intelligence (AI)

AI helps to optimise logistics operations by predicting demand and supply patterns in organisations. Organisations may rely on AI to help them in decision-making relating to inventories/stock. AI provides the ability for business to grow and gain competitive advantage in a cost-efficient way. "As costs come down and use cases become clear, it's providing both efficiencies and opportunities" (Klumpp, 2018; Embia et al., 2023b). As a way of practicing sustainable development, in terms of logistics activities in PNG, when engaging in economic activities, such as demand focusing, IBM stated that, "While no one can foresee what's in store for tomorrow, we can work today on building a 'smarter' global supply chain". AI makes it easier through its predictive capabilities that businesses earn less when inventories are low, because they can't supply what is demanded by customers. AI makes planning viable on network and demand, without businesses running out of merchandise and also contributes greatly towards reducing costs, thus saving time, replacing human activities in companies as well as reducing carbon emissions from harming the environment as businesses expand their activities through digital technology in (e-commerce) operation. This has boosted productivity and improved efficiency. For the logistics industry in PNG, the companies working and collaborating with technology providers is a way forward to drive future innovation and competitive advantage. Most logistic companies have progressed well when using AI by systemising their operation in simplifying the process involving logistics. A clear case is the UPS development On-Road Integrated Optimisation and Navigation Technology (ORION), which since 2012, has contributed greatly to the environment by saving ten million gallons of gasoline that could have been used on 1 hundred million miles (Horner, 2016). Hence AI delivers a sustainable development practice to companies when they perform their business operation in terms of distributing merchandise to retailers or customers.

Logistics companies are now embracing the use of digital technology as it delivers on time without much delay, and are convenient and trustworthy. Doing business while valuing the environment using innovative software programs on algorithms, artificial intelligence, and machine learning offers precise delivery time estimates, dependability, and responsiveness.

12.5.2 Internet of Things (IoT)

In general, when sensors are attached to products from suppliers, this allows to track the movement of the product using either mobile phones or laptops and computers, so long as there is connection to the network so that data can be exchanged over the internet. That is IoT. IoT may help to track goods and monitor environmental conditions during transportation. When sensors are placed on a physical object or product, people or companies can monitor the movement of the purchased item over the internet, to which everything is programmed and human efforts are replaced by robots, thus boosting performance. Customers won't be let down as all the stock are in place. This explains the application of IoT, the use of advanced robotics, and the application of advanced analytics of big data in supply chain management (Ben-Daya et al., 2019; Muduli et al., 2021; Peter et al., 2022; Embia et al., 2023a).

PNG is still in its early stages of adopting to applying IoT, but there are some examples of IOT applications in various industries. Example, in the mining sectors in PNG, IoT sensors can be used to monitor equipment performances and optimise mining operations, improving safety, efficiency, and profitability. Organisations such as National Agricultural Research Institute and various agricultural research institutions use IoT sensors to monitor soil moisture, temperature, and nutrients levels, enabling farmers to optimise crop yields and reduce water usage. Some logistics companies in PNG use services such as an IoT- enabled GPS tracker to track and manage inventory and shipments, improving supply chain efficiency and reducing losses, such as when they purchase goods from overseas. As IoT adoption continues to grow in PNG, we can expect to see more innovative applications of this technology in the future. Supply chain management will shift to a new dimension as numerous developments are underway; this will have a big impact on the logistics supply chain because most rural areas globally have been growing in terms of development activities (Piso et al., 2023; Galgal et al., 2023). The same applies to PNG as well. There is money floating everywhere unlike before as people continue to engage in economic activities through agriculture (cash crop, farming, etc). This puts more burden on the environment, because the amount of carbon emissions will be high as people engage in socio-economic activities, creating more challenge to the logistics industry.

12.5.3 Big Data Analytics

Big data analytics help to analyse large amounts of data generated by logistics activities to identify inefficiencies and opportunities for improvements. This is when human efforts are replaced by robots in handling large inventories. That includes activities such as picking and packing of inventory in large boxes or pallets to be sent to the warehouse and put on ships to other destinations as requested from business companies. Hence, big data analytics may create flexibility on the use of increased vehicles in moving products along the network. It also improves optimisation of cross-company engagement in terms of product movement. How the network is set up is the concern of most businesses requirements as the network alone can be the best fit to what the business desires.

According to McKinsey and company, predictive analysis on demand either internally or externally, such as market trends, weather conditions, school holidays, and so on, forms the need for those services which are required in the approaches of forecasting. Those data as well as machine status data for spare parts demand will provide accurate forecast of customer demand (Alicke et al., 2017; Ivanov et al., 2019). For instance, Amazon has patent with shipping or logistics company, thus they ship their product to their customer before they actually place orders, which is possible through applying big data to analyse future forecast for the business. They identify the customers buying habits through big data analytics, which assist to predict future shipments to its customers. Companies may look into customers' new queries by the use of digital technology in the supply chain in regard to their supply as well as improving efficiency. They solve dual challenges by simply adapting digital technology in their operation.

The application of big data analytics is a form of digitalisation that makes it easier in terms of distributing products and is convenient for business and customers where they receive products in an accurate time period. Big data analytics also creates flexibility and efficiency for businesses when products are distributed to them from suppliers. To avoid high truck involvement with distinct offers during delivery periods, products are distributed in the supply chain by following the tactics of dynamic planning, which well suits those activities which are shaped by higher demand for specific products. Two key enablers that allow the change in supply chain to become more digitalised are regarded as capabilities and environment. Employees in the company have to be trained well so they are capable of applying big data analytics in the supply chain process through logistics operation, including a better internet connectivity to speed up the supply chain process. This saves the environment from several harms as good are distributed, and thus promotes sustainability practices with reference to PNG context. Figure 12.4 provides a clear understanding of implementing digital technologies in sustainable practice in logistics at PNG and the country will get further benefits in long run.

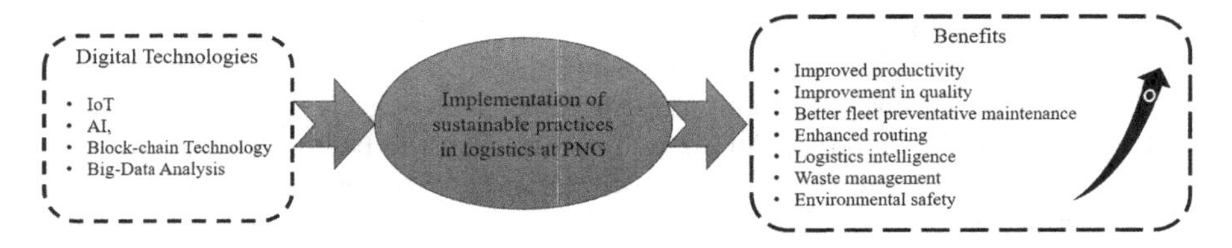

FIGURE 12.4 Link between Employing of Digital Technologies in Promoting a Sustainable Practice in Logistics and Further Benefits.

12.5.4 Opportunities

The opportunities PNG can tap into in terms of sustainability practice in Logistics is through the use of adapting digital technology, and that will become a bonus for the overall company itself and the environmental needs of the entire population in the country to safeguard the resources that we have that will continue to be used by the next generation and beyond. Sustainable logistics and digital technology are two important aspects of modern supply chain management that can help reduce environmental impact and improve efficiency. According to a report by the World Economic Forum, digital traceability can make supply chains more sustainable. Using AI technologies and IoT, manufacturers will feel that they have a duty to contribute to the environment by being responsible for the damage they are causing as in the case of a product development such as car batteries, phones, and so forth, apart from data analytics use of supply chain networks for logistics purpose.

Logistics has a massive influence in paving the way for businesses and government to see the need for adapting digital technologies to supply chain management in the flow of materials, either internal or external, following new processes with its activity, including new data providing information to acquire customers. The activities of the logistics sector in 2020 have added to a partial carbon dioxide emmision of 21% alone compared to other business sectors in operation around the globe. As customer demand increases, there are also pressures by regulators and investors with new green laws for the supply chain; as such, companies are required to engage in a sustainable logistics practice, which is eco-friendlier and more efficient (Maersk, 2022; Junge, 2019; Embia et al., 2023c; Chhotaray et al., 2024).

12.5.5 Challenges

The use of data analytics will help companies in making concrete decisions, which cuts down costs of transporting of goods, improves efficiency, and saves time. By applying data analytics, this allows logistics or transportation companies operating globally to increase their profit margins. In that way, they can solve the issues of managing costs in their business. PNG may also apply the same idea in applying data analytics to minimise costs when engaging in business within the country. Now the country has plans to increase its national road network over the next three years. When roads are connected to provinces, this will definitely increase economic activity, thus causing severe environmental damages as already several environmental risks have been identified when assessing the "national road networking plan" by analysing environmental and biophysical data via a fine scale. However, logistics companies operating in the country may find that costs will be higher for them when engaging in transportation of goods via the road link. At the same time, they may add more destruction to the environment from the cars' exhaust. To avoid all these environmental externalities and higher transportation costs, they can apply data analytics to overcome these challenges.

Looking at infrastructural development on the national road network in the country, though it may spur economic growth, the expansion of infrastructure will pose challenges to sustainable development in PNG. Though PNG houses the third largest tropical rainforest on the planet, we have to be mindful of our actions and decisions and what effects they may have on the environment. The challenge today in PNG is the implementation of green logistics. Application of an eco-efficient supply chain management in the flow of products, either backward or forward, and the information from production handing a ready product to consumers is one of the aspects of sustainable development to be considered and that defines green logistics as well. Therefore, the need for considering applying green logistics in PNG, particularly for economic and social development, will be a great challenge without fully exploiting all our natural resources. The sustainable development goal of green logistics is having a minimum environmental constraint by adapting to a digitalised supply chain management system. Also, sustainable development management is another challenge for PNG. Currently, we need to consider its implementation and application seriously so that we may minimise the risk of environmental damage.

12.5.6 Suggestions to Improve

It is challenging to set up and apply sustainable progress due to constant change in the growing technology with a vibrant worldwide economy. The PNG government's laws, acts, and policies need to be tightened for companies to follow; careful review and better strategies need put in place on how to execute them. These will set the benchmark for companies to follow according to the rules and guidelines as per the government requirement. Such effective control measures are a way to overcome the challenges. Through increased collaboration with partnership, this will capitalise on better management and green logistics implementation. In building good relationships both within companies and across countries, it is possible for stakeholders to win an agreement, and indeed this helps to create reasonable partnerships and collaborations between PNG and other countries.

Logistics firms need to ensure their leverage by using advanced analytics on performance management, fleet management, optimisation of logistics and supply chain, cost management, business intelligence, better customer services, and study of outcome of analysis. It is compulsory to comply with the principles of sustainable development, and this can be done by companies through communicating with their respective employees, business associates, and clients.

12.6 IMPLICATIONS

The implication of this study, through its results, will be helpful for industry people, academics, and researchers through the administering of an ecological process such as managing a sustainable development practice and will contribute to the achievement of the United Nations Sustainable Development Goals, the 2030 Agenda, and Vision 2050, which links UN Development Goals. With the need for adapting a pleasant-sounding climate worldwide, this coincides with the United Nations 2030 Agenda. For the world to be wealthy, the entire human race needs to preserve the environment and put energies into restoring resources. Gender equality among humans with chances of entrée in education and earning incomes leads to promoting people's welfare, and collaboration and partnering helps aid sustainable development enormously.

PNG, along with other countries of the world (UN Nations), through their active participation and safe practice of sustainable development in logistics, with proper management of a sustainable development practice by everyone may assist to soar to meet the required goals. Every country should adapt to changing technology and incorporate it when engaging in business activities all throughout the world.

12.7 CONCLUDING REMARKS

The application of data analytics is seen as the key driver and is important for those countries to stay alive and flourish in the dynamic business environment. Over the last decade, PNG had realised the significance of data analytics due to the global boom of e-commerce. Thus, data analytics, including predictive analysis, enables e-commerce business to continue on when engaging in reverse logistics and last-miles delivery. Normally, it is done through the real-time data collected for the purpose of the optimising of routes.

For most companies who are planning to rationalise their operation, including optimising the allocation of resources, managing fleet movements, and rerouting, the use of data analytics will assist them to carry out those activities required. Most logistics companies, for instance, DHL, are moving to a succeeding technology known as augmented reality. Those businesses that are in operation in PNG, may gain a competitive advantage over their difficulties through applying visualisation of data, predictive analytics and big data, and large supply chain management.

In order to have a workable benefit, companies are required to alter their supply chains through cost-saving measures, avoiding inefficiency, managing inventory, and improving preparation and projecting. For engagement contracts with a longer-term duration with several programs, data analytics makes it easier to adequately discuss costing and recommend answers with concerned suppliers. Hence, careful planning and implementation of a sustainable management system creates a prosperous world, creating agreement among countries. Also, it pertains to the entire planet's ecosystem, including humans. Management of sustainability paves the way to addressing issues such as change of weather patterns, exhausting of non-renewable natural resources, and ruining of the environment, which may affect it. Those issues are caused by human actions and decisions. Sustainability management through partnership and collaborations leads to preservation of resources – that is sustainable development where we humans use resources for ourselves and future generations can use them as well. The availability of resources has posed a great concern to humans as it is becoming less and less obtainable due to the growing population and emerging technologies. The resources need to be refurbished and this can be done by us, through a single individual, the society, the country, and the world through our actions and decisions when interacting responsibly to the environment. Administering an ecological progress is prerequisite to enforce the sustainability practice in supply chain management (logistics), capitalising on digital technologies. Undeniably, it is every inhabitant of the world who may give within a company to embrace an added eco-responsible approach.

REFERENCES

Alicke, K., Rexhausen, D. and Seyfert, A., 2017. Supply Chain 4.0 in consumer goods. *Mckinsey & Company*, *1*(11), 1–11.

Almeida, H.A. and Correia, M.S., 2016. Sustainable impact evaluation of support structures in the production of extrusion-based parts. In Muthu, S., & Savalani, M. (Eds.). *Handbook of Sustainability in Additive Manufacturing. Environmental Footprints and Eco-design of Products and Processes* (Vol. 1, pp. 7–30). Springer.

Ben-Daya, M., Hassini, E. and Bahroun, Z., 2019. Internet of things and supply chain management: a literature review. *International Journal of Production Research*, *57*(15–16), pp. 4719–4742.

Burton, I., 1987. Report on reports: Our common future: The world commission on environment and development. *Environment: Science and Policy for Sustainable Development*, *29*(5), pp. 25–29.

Centobelli, P., Cerchione, R. and Esposito, E., 2020. Pursuing supply chain sustainable development goals through the adoption of green practices and enabling technologies: A cross-country analysis of LSPs. *Technological Forecasting and Social Change*, *153*, p. 119920.

Chhotaray, P., Behera, B.C., Moharana, B.R., Muduli, K. and Sephyrin, F.T.R., 2024. Enhancement of manufacturing sector performance with the application of industrial Internet of Things (IIoT). In Behera, B.C., Moharana, B.R., Muduli, K., & Islam, S.M.N. (Eds.). *Smart Technologies for Improved Performance of Manufacturing Systems and Services* (pp. 1–19). CRC Press.

Cullinane, S., 2014. Greening logistics: Sustainable best practices, *SEStran document prepared in partnership with TRI.* https://sestran.gov.uk/wp-content/uploads/2017/01/action_4_task_2_sustainable_best_practices_final.pdf

Destiche, A., 2013. UPS saves millions and reduces CO_2 emissions with new ORION system. https://fieldlogix.com/news/ups-saves-millions-reduces-co2-emissions-new-orion-system/

Ehrhart, C.E. and Müller, J., 2010. *Delivering Tomorrow: Towards Sustainable Logistics.* Bonn: Deutsche Post AG.

Embia, G., Mohamed, A., Moharana, B.R. and Muduli, K., 2023a. Edge Computing-Based Conditional Monitoring. *Intelligent Manufacturing Management Systems: Operational Applications of Evolutionary Digital Technologies in Mechanical and Industrial Engineering*, pp. 249–270.

Embia, G., Moharana, B.R., Mohamed, A., Muduli, K. and Muhammad, N.B., 2023b. 3D printing pathways for sustainable manufacturing. In *New Horizons for Industry 4.0 in Modern Business* (pp. 253–272). Cham: Springer International Publishing.

Embia, G.J., Moharana, B.R., Behera, B.C., Mohmaed, N.H., Biswal, D.K. and Muduli, K., 2023c. Reliability prediction using machine learning approach. In *Smart Technologies for Improved Performance of Manufacturing Systems and Services* (pp. 21–37). CRC Press.

Evans, N.D., 2017. *Digital Sustainability: Digital Transformation's Next Big Opportunity.* https://www.cio.com/article/234278/digital-sustainability-digital-transformations-next-big-opportunity.html#:~:text=There's%20an%20opportunity%20for%20digital,in%20one%20form%20or%20another

Fröding, K. and Lawrence, G., 2010. Sustainability at IKEA. *Linnaeus Eco-Tech*, pp. 67–77.

Galgal, K.N., Ray, M., Moharana, B.R., Behera, B.C. and Muduli, K., 2023. Quality control in the era of IoT and automation in the context of developing nations. In Behera, B.C., Moharana, B.R., Muduli, K., & Islam, S.M.N. (Eds.). *Smart Technologies for Improved Performance of Manufacturing Systems and Services* (pp. 39–50). CRC Press.

Goundar, S., Purwar, A. and Singh, A., 2022. *Applications of Artificial Intelligence, Big Data and Internet of Things in Sustainable Development.* CRC Press.

Han, X., 2022, December. Analysis and Reflection of IKEA's Supply Chain Management. In Hrushikesh M., Gaikar V.B., & Ong T.S. (Eds.). *2022 4th International Conference on Economic Management and Cultural Industry (ICEMCI 2022)* (pp. 154–160). Atlantis Press.

Horner, P., 2016. Edelman award:'ORION'delivers success for UPS: revolutionary routing system boosts driver efficiency, cost savings, customer service and the environment. *OR/MS Today*, *43*(3), pp. 22–26.

IKEA Global, *Creating a sustainable IKEA value chain with IWAY.* https://www.ikea.com/global/en/our-business/people-planet/building-a-better-business-with-iway/

Ivanov, D., Tsipoulanidis, A., Schönberger, J., Ivanov, D., Tsipoulanidis, A. and Schönberger, J., 2019. Digital supply chain, smart operations and industry 4.0. *Global Supply Chain and Operations Management: A Decision-Oriented Introduction to the Creation of Value*, pp. 481–526.

Junge, A.L., 2019. Digital transformation technologies as an enabler for sustainable logistics and supply chain processes–an exploratory framework. *Brazilian Journal of Operations & Production Management*, *16*(3), pp. 462–472.

Kayikci, Y., 2018. Sustainability impact of digitization in logistics. *Procedia Manufacturing*, *21*, pp. 782–789.

Klumpp, M., 2018. Automation and artificial intelligence in business logistics systems: human reactions and collaboration requirements. *International Journal of Logistics Research and Applications*, *21*(3), pp. 224–242.

Larina, I.V., Larin, A.N., Kiriliuk, O. and Ingaldi, M., 2021. Green logistics-modern transportation process technology. *Production Engineering Archives*, *27*(3), pp. 184–190.

Laurin, F. and Fantazy, K., 2017. Sustainable supply chain management: a case study at IKEA. *Transnational Corporations Review*, *9*(4), pp. 309–318.

MAERSK, November 2022. Optimising digitalization for tangible sustainable logistics. https://www.maersk.com/news/articles/2022/11/15/optimising-digitalization-for-tangible-sustainable-logistics

Muduli, K. and Barve, A., 2013. Modelling the behavioural factors of green supply chain management implementation in mining industries in Indian scenario. *Asian Journal of Management Science and Applications*, *1*(1), pp. 26–49.

Muduli, K., Kusi-Sarpong, S., Yadav, D.K., Gupta, H. and Jabbour, C.J.C., 2021. An original assessment of the influence of soft dimensions on implementation of sustainability practices: implications for the thermal energy sector in fast growing economies. *Operations Management Research*, *14*, pp. 337–358.

Peter, O., Swain, S., Muduli, K. and Ramasamy, A., 2022. IoT in combating COVID-19 pandemics: lessons for developing countries. *Assessing COVID-19 and other pandemics and epidemics using computational modelling and data analysis*, pp. 113–131.

Piso, K., Mohamed, A., Moharana, B.R., Muduli, K. and Muhammad, N., 2023. Sustainable Manufacturing Practices through Additive Manufacturing: A Case Study on a Can-Making Manufacturer. *Intelligent Manufacturing Management Systems: Operational Applications of Evolutionary Digital Technologies in Mechanical and Industrial Engineering*, pp. 349–375.

Rodrigue, J.P., Slack, B. and Comtois, C., 2017. Green logistics. In *Handbook of logistics and supply-chain management* (Vol. 2, pp. 339–350). Emerald Group Publishing Limited.

Sandberg, E. and Abrahamsson, M., 2011. Logistics capabilities for sustainable competitive advantage. *International Journal of Logistics Research and Applications*, *14*(1), pp. 61–75.

Van Marwyk, K., Treppte, S. and Berger, R., 2016. Logistics Study on Digital Business Models. *Roland Berger, White Paper.*

Wailoni, X., Swain, S., Lafanama, S. and Muduli, K., 2022. Analytical approach for prioritizing waste management practices: implications for sustainable development exercises in healthcare sector. *International Journal of Social Ecology and Sustainable Development (IJSESD)*, *13*(1), pp. 1–12.

Wang, S., Wan, J., Li, D. and Zhang, C., 2016. Implementing smart factory of industrie 4.0: an outlook. *International Journal of Distributed Sensor Networks*, *12*(1), p. 3159805.

Wced, U., 1987. *Our Common Future-Brundtland Report* (No. UN Doc. A/42/427). United Nations World Commission on Environment and Development.

Wilson J., September 2020. Artificial intelligence in logistics: How AI can make your processes more efficient, Technology & Innovation. https://www.sage.com/en-gb/blog/artificial-intelligence-in-logistics-efficient-processes/

Yang, M., Fu, M. and Zhang, Z., 2021. The adoption of digital technologies in supply chains: Drivers, process and impact. *Technological Forecasting and Social Change*, *169*, p. 120795.

Yontar, E., 2022. Assessment of the logistics activities with a structural model on the basis of improvement of sustainability performance. *Environmental Science and Pollution Research*, *29*(45), pp. 68904–68922.

13 Advertising in the Era of Artificial Intelligence (AI)

An Opportunity for Circular Economy

Prabhuram Tripathy
Sri Sri University, Odisha, India

13.1 INTRODUCTION

Nearly every element of human life has been transformed by technological advancement over the years, including how businesses sell their products. A few emerging technologies, including robotics, big data analytics, artificial intelligence (AI), and the Internet of Things (IoT), are having a huge influence because of well-established breakthroughs like the Internet, mobile apps, and social media on how marketing is done by providing digital strategies for expanding and maintaining customers (Hoffman & Novak 2018, Mende et al., 2019, Bolton et al., 2018, Davenport & Ronanki 2018). Many different sectors have implemented AI in order to profit from any prospective advantages. Commercial houses are adopting AI and other cutting-edge technology as they get closer to Industry 5.0. (Chintalapati & Pandey, 2022). Advertising has shifted from conventional methods like radio, television, billboards, and newspapers to a variety of novel and intriguing media and platforms as a result of technological advances and the growth of digital media. Advanced media for advertising rely on the use of AI to increase the effectiveness of ads and simplify the delivery of advertisements. According to Qin and Jiang (2019) and Copeland (2021), AI is a group of innovative technologies that give computers the ability to make decisions, solve problems, and complete tasks that would typically require people and their brains. Many marketers are using AI to transform this vast volume of data into relevant customer information. Of course, there are dangers, as the results of Cambridge Analytica's historical analysis show that millions of Facebook profiles are being used for political objectives (Solon and Laughland, 2018).

13.2 ARTIFICIAL INTELLIGENCE

The scientific field of artificial intelligence aims to build computers and other devices that possess the ability to reason, learn, and do tasks that typically require cognitive ability or are too complicated for humans to easily understand. A handful of the numerous fields that fall under this broad area include computer science, statistics, data analytics, linguistics, neurology, philosophy, hardware and software

DOI: 10.1201/9781003349877-13

engineering, and psychology. AI is a phrase used to describe technologies that allow applications like machine learning and deep learning, which are the main foundations for forecasts, intelligent data retrieval, natural language processing, object classification, data analytics, and suggestions (Haleem et al., 2022).

The definition of artificial intelligence refers to computer programmes that gather and utilise data to do tasks in the best possible way that intelligent beings would. IBM's AI research director, Guruduth Banavar, refers to the field as "a portfolio of technologies" since there are so many different kinds of AI (Kaput, 2016). These AI technologies are being developed at varied rates and have a variety of purposes, but they are all concentrated on making computers "smart" by emulating human intelligence. As per Buch and Thakkar (2021), AI comes in two types, described in the following subsections.

13.2.1 Artificial General Intelligence (AGI)

Strong artificial intelligence, often known as artificial general intelligence (AGI), is a subset of AI and is capable of carrying out every task that can be completed by a rational creature. This kind of AI, which can do an extensive range of tasks and perform like humans, is commonly depicted in science fiction. True AGI has never been successfully created because of the complexity of human psychology and our limited understanding of how our minds work. For the remaining parts of this research, AGI will virtually go unnoticed since it has no real-world use for marketers at this time.

13.2.2 Narrow AI

Narrow artificial intelligence is commonly referred to as weak AI, yet it may perform some tasks. It emphases enhancing a specific set of cognitive abilities, including customer segmentation, driving, picture recognition, and predictive analysis. The most prevalent type of AI, which we encounter every day, is in the form of spam email bots and website recommendation algorithms, such as Netflix's personalised movie and TV programme recommendations and Amazon's suggested purchase ideas.

13.3 MODERNIZATION OF ADVERTISING WITH TECHNOLOGY

Since advertising has been around for a while, both marketing and consumer psychology have collectively substantial study literature (Yadati et al., 2013). Artificial intelligence applications in current advertising are significant. AI is the capability of a computer programme, computer-controlled devices, or robot to do tasks usually performed by people with intelligence. The word is typically used to describe efforts to develop and advance human-like intellectual and reasoning systems, such as those that can reason, grasp, find meaning, discern, generalise, or learn from the past. A subset or component of AI is machine learning. It's a broad idea that machines might be able to execute.

The use of modern technology is widespread in advertising. Big data provides digital and smart marketers with vital information and insights about their target audiences. Big data analytics services on a large scale now enable businesses to collect, store, and evaluate both structured and unstructured collected data. Based on the obtained Big data, access behaviours, profiles, and movement patterns of mobile users, advertising suggestions are made using the stated analytics approaches. Both online and offline advertising activities are made possible by big data (Jin et al., 2015), but it has a number of problems, including invasion of client privacy, an unstructured data pattern, the inability to directly provide findings at the user level, the difficulty of transferring data easily, and the manipulation of customer information.

The shortcomings of the aforementioned technologies can be solved through machine learning, a branch of AI and one of the most fascinating technologies I've ever encountered. The large field of machine learning in science and technology enables computers to acquire knowledge without being largely programmed. Due to the simultaneous availability of huge quantities of real data on consumer behaviour, information, addressing the brand-focused activities of the buyer through the buying organisation approach, and the capability to both plan and execute advertising campaigns in real time, the complex ecosystem necessitates that machine learning take the lead in the ad escalation process (Shah et al., 2020).

13.4 USE OF AI IN PROMOTION MANAGEMENT

The term "promotion" in the context of the marketing mix describes the process of informing the target market about a good, service, brand, or activity in order to attract and persuade them to purchase it. Media planning, media scheduling, advertising management, and other aspects of promotion management are included. With the advent of digital marketing, company promotional techniques are moving from traditional to digital platforms; social media marketing is the most often employed strategy. Customer preferences govern the time, location, and content of digital campaigns. According to the client profile and preferences, AI enables the modification and personalization of messaging (Huang & Rust, 2021). Real-time client likes and dislikes may be tracked using emotional AI-based algorithms. Content analytics may increase the usefulness and impact of messages. Thanks to social media information based on netnography, advertisers now have more alternatives for personalising their marketing campaigns to fit the interests of their target market (Verma, 2018; Verma & Yadav, 2021).

13.5 MOTIVATING FACTORS FOR AI ADOPTION IN ADVERTISING

The development of intelligent advertising seems to be fuelled by big data, algorithms, and cloud computing, whereas programmatic and interactive advertising were created for a number of technological, economic, and social reasons. However, a variety of factors, such as use in the field, industry innovation, governmental regulation, business acceptability, and end-user acceptance, may have an effect on its future development. By carefully examining these issues, we will certainly be able to better understand the direction smart advertising is taking both domestically and globally (Li, 2019; Kietzmann and Canhoto, 2013).

13.5.1 Artificial Intelligence in Advertising Is a Feature of the New Era of Advertising

It was made evident by James Cannella (2018) that the era of marketing with AI is swiftly arriving and has far-reaching ramifications. As AI rapidly evolves and is widely used in marketing, the capability of marketers to utilise and manage AI solutions will become an increasingly important skill set. AI still benefits marketers, consumers, and society at large because it makes it simpler for advertisers to generate and provide the appropriate individuals with value at scale in the right manner at the right time. This is true even with the considerable difficulties that need to be handled before widespread acceptance. As repetitive jobs are automated by AI, marketers will be able to concentrate more of their attention on tasks that improve customers' lives, increase the rate at which workplace goals are completed, and promote creative thinking for the benefit of both customers and sellers. From James Cannella's (2018) findings, artificial intelligence in the era of marketing will fundamentally alter how marketers connect with customers and how they accomplish their objectives. In the digital age, AI is quickly advancing, and the business sectors of marketing and advertising are not an exception.

The relationship between digital marketing and AI is discussed in this chapter, and the former is seen as the field's most promising future. Both emphasised how changing tactics and techniques will alter the marketing environment. Additionally, it alters how marketers execute their efforts, including how they are evaluated and managed. The future and current practices of digital marketing will be governed by AI. Artificial intelligence will thus be making waves in the online world. According to Davenport et al. (2020), AI has a significant impact on the advertising and marketing sectors, where it is emphasised that the use of artificial intelligence in marketing strategies is on the rise. AI has the potential to improve yield management and provide exceptional customer service by implementing dynamic pricing. Additionally, Davenport et al. displayed the analysis of the information gathered from ten different Pakistani organisations. The highly recognised marketing management support system uses AI to empower managers to make quick choices and to assess and present data and information.

13.5.2 AI in Advertising Today

Of marketers who showed curiosity about using AI, it is expected that 98% will start to use AI in the near future. Only 20% of them adopted one or more AI technologies at scale as a core component of their business processes in 2017. Though some may think marketers are already behind due to the spike in interest in the subject, this discrepancy between excitement and execution shows that it is still possible for them to use AI. Nevertheless, in addition to the vast array of front-line tools and services now available for businesses to utilise, uses of AI in marketing are continually developing. This indicates that the usage of AI in marketing will reach a turning point in 2018 and the following 24 to 48 months. Marketing is now the fourth-largest use case in terms of resources invested and the sixth-largest industrial user of AI technology, accounting for 2.55% of the industry's total investment in AI (Arsenault, 2015). Although AI has been utilised in marketing for many years, a variety of issues have backed the current rise in attention and usefulness. These include an explosion of interest in the sector, a growing pool of highly qualified experts eager to progress the business, and big data and the improvements in data management that came along with it. These include, but are not limited to, cheaper prices than ever before and improved computer capacity to process AI algorithms at scale. With AI technology becoming more widely available than ever, its marketing potential is beginning to take shape in a variety of ways. Although there are currently very few companies adopting robust AI systems, many companies of all kinds are employing small-scale solutions that are easier to set up and sustain. The sum of money essential to develop and continue AI systems, the extent to which AI is essential to their fundamental business propositions or daily operations, as well as the complications of their AI applications, should be considered. Reduced entry barriers (such as fewer resources needed) benefit businesses that use low-involvement AI solutions, but as will be shown in the following, they might not be able to fully take advantage of the benefits that more involved, comprehensive AI systems have to offer in terms of setup and maintenance. On the other hand, using high-involvement AI solutions by organisations may greatly raise resource requirements while having many beneficial advantages for their core business operations (Buch & Thakkar, 2021; Schiessl, et al., 2021).

13.5.3 AI's Effect on the Advertising Sector

Although Qin and Jiang (2019) contend that the primary effect of AI on the advertising process has been to boost efficiency, the effect of AI on the industry's organisational structure is still an unresolved subject. The connection between clients and advertising agencies has evolved in several respects as a result of the introduction of programmatic advertising. For instance, some consulting firms help their customers create internal resources to manage particular tasks like media buying and campaign research. Such structural changes may accelerate as more clients employ AI technology in advertisement development, which would be detrimental to the agency sector (Li, 2019). AI is becoming more and more popular, and as a result, marketing research is focusing on it more. Marketers are using AI to improve their marketing and advertising efforts and successfully reach their target audience as advertising technology advances.

13.5.4 Impact AI Has on Advertising along the Consumer Journey

To properly comprehend the possibilities AI brings for marketers, one must first recognize how communications typically "work" throughout the customer's decision-making process. The client must first understand their demands in order to go through the phases of initial thought, active evaluation, purchase, and post-purchase (Court et al., 2009). In this part, Kietzmann et al. (2018) outline the consumer processing procedures, advertising goals, and customary advertising duties for marketers at each stage of the customer journey. They then demonstrate how these advertising jobs are transformed by the aforementioned AI-building elements (2018).

13.6 FUTURE SCOPE AND CHALLENGES

Despite the fact that this chapter was written with excellent precision and thorough study, there are several difficulties that might affect how this work develops in the future. Business and society are increasingly being impacted by artificial intelligence, and machine learning presents challenges for the advertising industry on both a practical and philosophical level. The main problem with the use of machine learning and AI is that errors may occur in the algorithms, and their implementation may lead to significant issues. The key challenge in machine learning is acquisition. The data must be processed using a variety of algorithms. Before being fed into the right algorithms, data must also be processed. Therefore, it has a big impact on the results. To accomplish some AI activities, such as optimising, customising the material for the target audience, and so forth, we now have to rely on individual solutions. Another significant issue with machine learning is interpretability, which is a crucial attribute that machine learning techniques should strive for if they are to be used in real-world applications. These are a few crucial issues that warrant consideration and additional research (Shah et al., 2020).

AI has a huge future in marketing. Advertising may alter significantly by utilising algorithms for targeted display advertising. Using machine learning algorithms may have a significant influence on how well a marketing strategy performs overall for both small and large businesses. The campaign is made more noteworthy and inventive with the aid of predictive analytics tools. From a business producing nanofibers to massive machinery, techniques using AI and machine learning assist in concentrating on the intended clientele and providing successful outcomes through digital advertising. Additionally, it will significantly alter both the customers' perception of digital advertising as well as the marketing methods used by businesses. For their advertising budgets, businesses will spend more on AI and machine learning technology.

13.7 CONCLUSION

The objective of advertising is to convince someone in particular to purchase a good or service. This chapter offers a framework for comprehending the potential future effects of AI and machine learning on advertising, with a focus on how these technologies will control consumer behaviour and marketing strategies. Even if there is a lot more technology nowadays, such as cloud computing, big data, and so on, thanks to technical improvements, machine learning and AI have outperformed other technologies in terms of effectiveness. AI and machine learning improve business marketing techniques. The implementation of machine learning and AI techniques will make it much simpler to design innovative ads, target advertising, and optimise performance than it was in the past. However, there are a few challenges that AI and machine learning in advertising face, but they should be overcome shortly by using machine learning and AI to monitor the outcomes. AI has altered how advertising understands and directs consumers. New user-generated data mining techniques will be the driving force behind future consumer insights, with AI acting as the privacy standard. By employing machine learning, advertisers will be able to covertly collect customer information from several sources, consolidate it, and evaluate it to provide actual consumer insights.

REFERENCES

Arsenault, M. "Remarketing vs. Retargeting: What's the difference." Retrieved April 5 (2015): 2018.

Bolton, Ruth N., Janet R. McColl-Kennedy, Lilliemay Cheung, Andrew Gallan, Chiara Orsingher, Lars Witell, and Mohamed Zaki. "Customer experience challenges: bringing together digital, physical and social realms." *Journal of Service Management* 29, no. 5 (2018): 776–808.

Buch, Ishaan, and Maher Thakkar. *AI in Advertising*. (2021): 1–18.

Cannella, J. Artificial intelligence in marketing. Honors Thesis for Barrett, The Honors College at Arizona State University, (2018): 1–132.

Chintalapati, Srikrishna, and Shivendra Kumar Pandey. "Artificial intelligence in marketing: a systematic literature review." *International Journal of Market Research* 64, no. 1 (2022): 38–68.

Copeland, B. J. Encyclopedia Britannica: Artificial Intelligence. (2021). December.

Court, D., D. Elzinga, S. Mulder, and O. J. Vetvik. "The consumer decision journey." *McKinsey Quarterly*, June 2009.

Davenport, Thomas, Abhijit Guha, Dhruv Grewal, and Timna Bressgott. "How artificial intelligence will change the future of marketing." *Journal of the Academy of Marketing Science* 48 (2020): 24–42.

Davenport, Thomas H., and Rajeev Ronanki. "Artificial intelligence for the real world." *Harvard Business Review* 96, no. 1 (2018): 108–116.

Haleem, Abid, Mohd Javaid, Mohd Asim Qadri, Ravi Pratap Singh, and Rajiv Suman. "Artificial intelligence (AI) applications for marketing: A literature-based study." *International Journal of Intelligent Networks* 3 (2022): 119–132.

Hoffman, Donna L., and Thomas P. Novak. "Consumer and object experience in the internet of things: an assemblage theory approach." *Journal of Consumer Research* 44, no. 6 (2018): 1178–1204.

Huang, Ming-Hui, and Roland T. Rust. "A strategic framework for artificial intelligence in marketing." *Journal of the Academy of Marketing Science* 49 (2021): 30–50.

Jin, Songchang, Wangqun Lin, Hong Yin, Shuqiang Yang, Aiping Li, and Bo Deng. "Community structure mining in big data social media networks with MapReduce." *Cluster Computing* 18 (2015): 999–1010.

Kaput, Mike. "The Marketer's guide to artificial intelligence terminology." Retrieved December 16 (2016): 2020.

Kietzmann, J., and A. Canhoto. "Bittersweet! Understanding and managing electronic word of mouth." *Journal of Public Affairs* 13, no. 2 (2013): 146–159.

Kietzmann, Jan, Jeannette Paschen, and Emily Treen. "Artificial intelligence in advertising: How marketers can leverage artificial intelligence along the consumer journey." *Journal of Advertising Research* 58, no. 3 (2018): 263–267.

Li, Hairong. "Special section introduction: Artificial intelligence and advertising." *Journal of advertising* 48, no. 4 (2019): 333–337.

Mende, Martin, Maura L. Scott, Jenny van Doorn, Dhruv Grewal, and Ilana Shanks. "Service robots rising: how humanoid robots influence service experiences and elicit compensatory consumer responses." *Journal of Marketing Research* 56, no. 4 (2019): 535–556.

Qin, Xuebing, and Zhibin Jiang. "The impact of AI on the advertising process: The Chinese experience." *Journal of Advertising* 48, no. 4 (2019): 338–346.

Schiessl, D., H.B.A. Dias, J.C. Korelo, "Artificial Intelligence in marketing: a network analysis and Future Agenda." *Journal of Marketing Analytics* (2021): 1–12.

Shah, Neil, Sarth Engineer, Nandish Bhagat, Hirwa Chauhan, and Manan Shah. "Research trends on the usage of machine learning and artificial intelligence in advertising." *Augmented Human Research* 5 (2020): 1–15.

Solon, O., and O. Laughland. "Cambridge analytica closing after Facebook Data Harvesting Scandal." *The Guardian*, May 2, 2018.

Verma, Mudit. "Artificial intelligence and its scope in different areas with special reference to the field of education." *Online Submission* 3, no. 1 (2018): 5–10.

Verma, S. and Yadav, N. "Past, present, and future of electronic word of mouth (EWOM)." *Journal of Interactive Marketing* 53, no. 1 (2021): 111–128.

Yadati, Karthik, Harish Katti, and Mohan Kankanhalli. "CAVVA: Computational affective video-in-video advertising." *IEEE Transactions on Multimedia* 16, no. 1 (2013): 15–23.

14 Minimum Quantity Lubrication
An Alternative Sustainable Cooling Technique for Machining Industry

Bikash Chandra Behera
C. V. Raman Global University, Bhubaneswar, India

14.1 INTRODUCTION

Machining Industries today face significant challenges when machining hard-to-cut materials like titanium, nickel, and ceramics. The machining of hard-to-cut materials presents significant challenges, including elevated cutting forces, increased vibrations in machining systems, concentrated heat generation, rising cutting temperatures, accelerated tool wear, and the potential for catastrophic tool failure.

To tackle these challenges, cutting fluids are mainly used, but conventional options pose problems due to high risk and cost. Flood cutting fluids, despite their widespread use in metal cutting, can lead to machine tool damage, health risks for operators, and disposal costs. Despite these challenges, the metal cutting industry still heavily relies on flood coolants. However, growing environmental concerns and associated financial penalties are driving research into cost-effective alternatives.

An effective coolant should demonstrate superior thermal properties, including low surface tension, high thermal properties, and the ability to maintain these properties at high temperatures. The lubrication capability is contingent on tribological properties such as viscosity, wear resistance, and corrosion resistance (Brinksmeier et al. 2015; Behera, Alemayehu, et al. 2017a; Behera, Chetan, et al. 2017b; Chetan and Venkateswara Rao 2015). The cutting fluid also serves to prevent the formation of built-up edge when applied to the cutting zone (Behera, Chetan, et al. 2017b). Although machining performance enhances due to the application of cutting fluids, the use of traditional cutting fluids in the industry has documented adverse effects on the environment, health, and overall machining costs.

To achieve proper cooling and lubrication in the machining zone without sacrificing the environment, various cooling techniques have been recently developed. However, this discussion specifically focuses on the recently developed minimum quantity lubrication (MQL) machining environments. The current chapter discusses various types of cutting fluids, their benefits, challenges associated with cutting

DOI: 10.1201/9781003349877-14

fluids, and alternative cooling and lubrication techniques and provides detailed insights into the MQL process.

14.2 TYPES OF METAL WORKING FLUIDS

In machining, different situations need specific ways to cool and lubricate. The needs for cooling and lubrication in machining can be different, so special fluids are made to fit each case. These fluids usually have a basic liquid with extra stuff that stops wear. They can be gases, mixtures of gas and liquid, regular liquids, mixtures of liquid with added things, or pastes (Heisel et al., 1994; Brinksmeier et al., 1999).

When we're doing precise and finishing machining, we often use cutting oils to make sure machining goes smoothly. But using these oils needs careful steps and special tools because they can make mist and fumes, and they might even cause a fire. Also, we have to think about pollution and safety when using cutting oils.

Metalworking fluids (MWFs) are really important in machining. There are five types based on what they're made of: synthetics, semisynthetic, soluble oil, straight (or neat) oil, and nano fluids. We pick a specific type of MWF by looking at what each one is good at and what might be a problem. Table 14.1 has details about the good and not-so-good parts that help us decide which MWF is best for the cutting.

TABLE 14.1
Benefits and Drawbacks of Metal Working Fluids (Byers 2016)

Metal Working Fluids	Advantages	Disadvantages
Synthetic	Efficient heat transfer with minimal environmental impact.	Leads to a reduction in the lifespan of the tool and causes corrosion to the workpiece during prolonged contact.
Semi-Synthetic	Effective lubricating properties suitable for medium to heavy-duty machining operations.	Demand high-quality water but are prone to easy foaming.
Soluble oils	Offer improved resistance to corrosion and enhanced capability for penetrating cracks.	Lack stability in emulsion and are susceptible to effortless separation.
Neat oils	Adequate lubricity extends the lifespan of the wheel. Oils with higher viscosity exhibit stronger adherence, resulting in reduced misting.	Adverse environmental impact and elevated expenses associated with maintenance and disposal.
Nano Fluids	Nano fluids enhance heat transfer efficiency through increased surface area and thermal conductivity, making them valuable for cooling applications in machining. Their unique properties hold promise for improving the efficiency of thermal systems.	Particle agglomeration, heightened viscosity, cost, environmental concerns, and stability issues, posing challenges to their widespread use.

14.3 METALWORKING FLUID – CHALLENGES

The impact of MWFs on the working environment has become a growing concern addressed by increasing regulations. These concerns cover a range of issues, from the potential harm of ingesting particles from the workpiece material or the process fluid—possibly causing cancer—to the irritating effects of metals, oils, and bacteria when inhaled or in direct contact with the skin. The health of machine operators can be affected by various substances present in the cutting fluid. Exposure through skin contact, swallowing, or breathing in oil mist and vapors can impair their health. To address these problems, there's a growing trend towards enclosing the entire machine and using fume extraction systems. The debate also considers the advantages of oil versus water-based fluids and the preference for using the minimum amount of fluid (Lodhi, Kumar, and Ghosh 2023).

Most MWFs create an environment where bacteria can grow, posing a hazard to machine operators. These fluids are also known to cause skin disorders, such as dermatitis. Additionally, there's the potentially fatal risk of heavy metals leaching into the fluid, affecting the human respiratory and dietary systems (Chetan and Venkateswara Rao 2015).

After the fluid has been used, it contains small amounts of debris. Proper disposal of the oil is crucial due to its significant environmental hazard. The combination of increasing disposal costs due to new environmental laws and uncertainty about health risks to machine operators has led to more research aimed at reducing costs through recycling, reducing consumption, or even eliminating the fluid from the cutting process. Numerous research studies in the literature explore techniques like MQL, dry or near-dry machining, and cryogenic machining, especially in processes like turning, drilling, and milling. Nowadays, researchers are increasingly experimenting with these methods in machining processes as well.

14.4 ALTERNATIVE COOLING/LUBRICATION TECHNIQUES

Several alternative cooling and lubrication techniques are employed in machining processes to address environmental concerns, improve efficiency, and overcome challenges associated with traditional flood cooling (Chetan and Venkateswara Rao 2015; Sen et al. 2023; Hong 1999). Some examples are included in the following.

14.4.1 Minimum Quantity Lubrication (MQL)

MQL involves applying a minimal amount of lubricant directly to the cutting tool or workpiece, reducing fluid usage while maintaining effective lubrication (Behera et al. 2022).

14.4.2 Compressed Air Cooling

Compressed air cooling presents an effective alternative to traditional flood cooling. Specifically, air cooling, a subtype of gas cooling, utilizes dry compressed air as a substitute for cutting fluid. This method is regarded as a cost-effective cooling solution, given the ready availability of compressed air in various shop floors and factories.

14.4.3 Cryogenic Machining

Cryogenic machining involves the use of extremely cold cutting fluid which is below 0°C, often using liquid nitrogen, to cool the cutting zone. This method is effective in reducing heat and extending tool life (Behera et al. 2022).

14.4.4 Solid Lubricants

Solid lubricants, such as MOS_2, WS_2, and graphite can be applied to the cutting zone to reduce friction.

14.4.5 High-Pressure Jet Environment

Using a high-pressure cutting fluid jet at the chip-tool interface is known as HPJ machining. In this method, the high-pressure jet enters the chip-tool interface, reducing the length of contact between the chip and tool, and removes the heat generated at the contact zone. It also creates a hydraulic layer between the chip and tool, minimizing frictional force. This process is aimed at enhancing efficiency and reducing friction during machining (Behera et al. 2022).

14.5 MINIMUM QUANTITY LUBRICATION

Minimum Quantity Lubrication (MQL) is a way of using a tiny amount of cutting fluid for machining. Unlike traditional systems that use a lot of fluid, MQL typically uses only 10 to 500 ml per hour. In MQL, a tiny amount of metal working fluid is sheared by compressed air and sent to the cutting area through a nozzle as a mist. The pressure of air in an MQL is usually between 0.4 and 1.0 MPa. This oil and air mix acts as a coolant and lubricant in the cutting zone, reducing the heat caused by friction at chip-tool interface. The compressed air carries away some of the excess heat from the machining area. The schematic of MQL is shown in Figure 14.1.

14.5.1 Atomization Principle in MQL

Atomization is the method of breaking down a large quantity of fluid into tiny droplets. Several types of atomizers are employed for this purpose, including 'air-assisted' atomization, 'airless' atomization, centrifugal atomization, ultrasonic atomization, and electrostatic atomization. Various fluid properties, such as density, surface tension, and viscosity among others, play a role in influencing the atomization process. Details of atomization source, energy source, and principle are:

1. Atomization source: Airless
    - Energy source: Fluid pressure

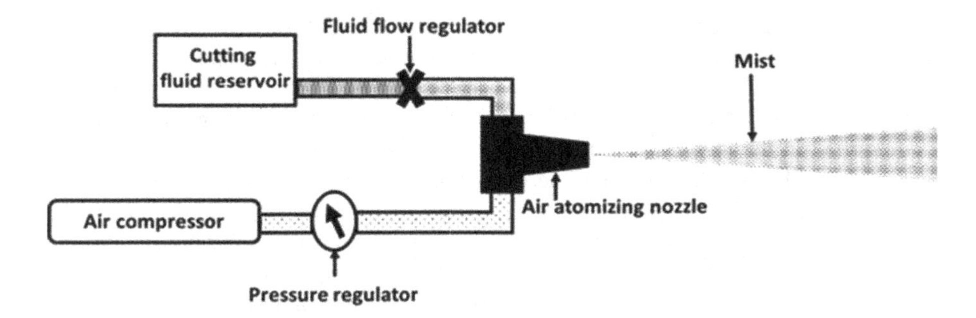

FIGURE 14.1 Schematic diagram of MQL setup (Behera 2018).

- Principle: The interaction between the liquid stream and air results in the atomization process.
- Factors influencing: The atomization process is influenced by factors such as nozzle design, fluid properties, and fluid pressure.

2. Atomization source: Air assisted
   - Energy source: Air pressure
   - Principle: The friction between the liquid and compressed air leads to atomization.
   - Factors influencing: Flow rate, fluid properties, nozzle design, and air pressure.
3. Atomization source: Centrifugal
   - Energy source: Centrifugal force
   - Principle: Liquid droplets are induced by the rotation of liquid over the spinning plate.
   - Factors influencing: Flow rate, fluid properties, and rotation of disc.
4. Atomization source: Electrostatic
   - Energy source: Electrostatic force
   - Principle: Atomization is brought about by the repulsive force between the atomizer and the liquid.
   - Factors influencing: Flow rate, fluid properties, electric field strength.
5. Atomization source: Ultrasonic
   - Energy source: Ultrasonic vibration
   - Principle: The fluid breaks into droplets due to vibration.
   - Factors influencing: Flow rate, fluid properties, amplitude of vibration, and frequency.

14.5.2 Classifications of MQL Systems

In the absence of an established classification for MQL, practitioners in the field, such as engineers and researchers, face considerable challenges in making informed decisions regarding the selection of suitable MQL equipment. A fundamental means of categorizing MQL systems is based on the methodology employed for delivering aerosols into the machining zone. The subsequent enumeration delineates various types of MQL systems that can be integrated into machining and grinding processes:

External aerosol supply: The aerosol is provided by an external nozzle positioned in the machine, akin to a nozzle for flood MWF supply. True to its name, MQL with an external aerosol supply incorporates the external nozzle responsible for dispensing the aerosol.

Ejector nozzle: The oil and compressed air get sent to the ejector nozzle, and the aerosol is made right after the nozzle. Basically, the nozzle acts as the atomizer, and it's the tool that helps control the ratios of air to oil by the way the nozzle is designed.

Internal (through-tool) aerosol supply: The mist is delivered through the tool, just like how high-pressure internal MWFs are supplied.

14.6 ECO-FRIENDLY AND ECONOMIC ASPECTS OF MQL

In MQL machining, a minimal quantity of cutting fluid is utilized, contrasting with traditional flood cooling. Consequently, there's no need for a reservoir to contain the cutting fluid. Conversely, flood cooling necessitates a reservoir for fluid storage. As the applied cutting fluid in the machining zone evaporates, there is no requirement for a storage tank. Consequently, the formation of bacteria is nearly negligible in MQL. The expenses associated with storing and recycling cutting fluid are unnecessary in MQL, resulting in a cost reduction of approximately 1000 times for cooling and lubrication. Moreover, expensive cutting fluids such as nano fluids and vegetable oil can be employed due to the reduced volume of cutting fluid required. MQL with vegetable oil serves as an environmentally conscious alternative for cooling and lubrication in machining industries. Its use aligns with the growing emphasis on sustainability and eco-friendly practices within the manufacturing sector. The MQL with vegetable oil is generally seen as safe for both the environment and humans. From an ecological and economic perspective, MQL is highly desirable.

14.7 SUMMARY

This chapter focuses on the difficulties faced by machining industries when dealing with hard-to-cut materials. These materials pose challenges due to their high strength at elevated temperatures, subpar thermal properties, and the existence of carbides in their microstructure. Traditionally, industries have relied on conventional cutting fluids like neat oil and soluble oil for machining processes. However, the use of these traditional fluids has been associated with various health and environmental issues. To address these concerns, there is a recent shift towards adopting alternative cooling and lubrication techniques. One notable method gaining attention is MQL. MQL presents itself as a promising solution by offering not only a cost-effective approach for fabrication but also an environmentally friendly method for the machining industry.

REFERENCES

Behera, Bikash Chandra. 2018. *Some investigations on the turning of inconel 718 and modeling for cutting forces* (Thesis).

Behera, Bikash Chandra, Habtamu Alemayehu, Sudarsan Ghosh, and P. Venkateswara Rao. 2017a. A comparative study of recent lubri-coolant strategies for turning of Ni-based superalloy. *Journal of Manufacturing Processes 30*:541–552.

Behera, Bikash Chandra, Chetan, Dinesh Setti, Sudarsan Ghosh, and P. Venkateswara Rao. 2017b. Spreadability studies of metal working fluids on tool surface and its impact on minimum amount cooling and lubrication turning. *Journal of Materials Processing Technology 244*:1–16.

Behera, Bikash Chandra, Chetan, Sudarsan Ghosh, and P. Venkateswara Rao. 2022. 2 - The underlying mechanisms of coolant contribution in the machining process. In *Machining and Tribology*, edited by A. Pramanik: Elsevier.

Brinksmeier, E., C. Heinzel, and M. Wittmann. 1999. Friction, cooling and lubrication in grinding. *CIRP Annals 48*(2):581–598.

Brinksmeier, E., D. Meyer, A. G. Huesmann-Cordes, and C. Herrmann. 2015. Metalworking fluids—Mechanisms and performance. *CIRP Annals 64*(2):605–628.

Byers, Jerry P. 2016. *Metalworking fluids*: CRC Press.

Chetan, Sudarsan Ghosh, and P. Venkateswara Rao. 2015. Application of sustainable techniques in metal cutting for enhanced machinability: A review. *Journal of Cleaner Production 100*:17–34.

Heisel, Uwe, Marcel Lutz, Dieter Spath, Robert A. Wassmer, and Ulrich Walter. 1994. Application of minimum quantity cooling lubrication technology in cutting processes. *Production Engineering 2*(1): 49–54.

Hong, Shane Y. 1999. Economical and ecological cryogenic machining. *Journal of Manufacturing Science and Engineering 123*(2):331–338.

Lodhi, Ajay Pratap Singh Deepak Kumar, and Sudarsan Ghosh. 2023. Performance evaluation of newly developed environmentally friendly metalworking fluids during grinding. *Biomass Conversion and Biorefinery*, 1–15.

Sen, Binayak, Shravan Kumar Yadav, Gaurav Kumar, Prithviraj Mukhopadhyay, and Sudarsan Ghosh. 2023. Performance of eco-benign lubricating/cooling mediums in machining of superalloys: A comprehensive review from the perspective of Triple Bottom Line theory. *Sustainable Materials and Technologies 35*:e00578.

Index

Pages in *italics* refer to figures and pages in **bold** refer to tables.

G

H

I

J

K

L

M

N

O

P

R

S

T

W

Printed in the United States
by Baker & Taylor Publisher Services